Nuclear models and the search for unity in nuclear physics

Krishna Kumar
Physics Department
Tennessee Technological
University

UNIVERSITETSFORLAGET
BERGEN · OSLO · STAVANGER · TROMSØ

© UNIVERSITETSFORLAGET 1984
ISBN 82-00-07149-9

Printed in Norway by
Combi-Trykk A/S, Bergen

Based on lectures
given at the University of Bergen,
Bergen, Norway, 1979-80

Distribution offices:
NORWAY
Universitetsforlaget
Postboks 2977, Tøyen
0608 Oslo 6

UNITED KINGDOM
Global Book Resources Ltd.
109 Great Russell Street
London WC1B 3NA

UNITED STATES and CANADA
Columbia University Press
136 South Broadway
Irvington-on-Hudson
New York, NY 10533

Contents

List of Figures
List of Tables
Preface

I. Introduction 1

  A. Historical Background 1
  B. Different Methods of Constructing A Many-Body Wave Function ... 7
     1. Particle-Hole Method 7
     2. j^n Method 7
     3. Quasi-Boson (RPA) Method 7
     4. Generator-Coordinate-Method 9
     5. Dynamic Deformation (Collective Hamiltonian) Method 9
  C. Main Approximations of the Dynamic Deformation Theory 10

II. Microscopic Theory of Collective Motion (How to go from the Time-Dependent-Schrödinger-Equation to the Colelctive-Schrödinger-Equation?) 13

III. Bohr's Collective Hamiltonian 19

  A. The Hamiltonian and its Symmetries 19
  B. Exact Treatment of Rotation-Vibration Coupling (Numerical Solution Method) 26
  C. Spherical (Vibrational) Limit 33
  D. Deformed (Rotational) Limit 46

IV. Deformed Single Particle Motion 56

  A. Rainwater-Nilsson Hamiltonian and its Parameters 56
  B. Scaling Method Leading to a Common Basis for all Nuclei 59

V. Pairing Correlations 63

  A. Improved BCS Theory Including the Particle-Hole Channel 63
  B. Determination of Global Parameters 70

VI. Potential Energy of Deformation 73

  A. Strutinsky Method 73
  B. The Droplet Model 77
  C. The Projection Correction 78

VII. Moments of Inertia and Mass Parameters 80

  A. Cranking Formulae 80
  B. Effects of Pair Fluctuations 82

VIII. Calculations of Electromagnetic Moments and Transition Probabilities 85

IX.	Discussions of Results (Dynamic Deformation Model)	88
	A. Parameters of the Model	88
	B. A Global Study of the Low-energy, Low-spin Properties of Even-Even Nuclei (A = 12-240)	88
	C. Shape Co-existence	88
	1. Shape Co-existence in ^{16}O	96
	2. Shape Co-existence in ^{72}Se	99
	3. Shape Co-existence in ^{240}Pu (Fission Isomers)	99
	D. Rotational-Vibrational Bands in Er Nuclei	102
X.	Conclusions	110
	Bibliography	112

List of Figures

		Page
1.	Projections of nuclear shapes	2
2.	Schematic pictures of the nuclear shell model and the oscillator model	3
3.	Potential energy vs. deformation	5
4.	Schematic pictures of the nuclear rotational model and the rotational spectrum	6
5.	Schematic picture of the particle-hole method	8
6.	Schematic picture of the equivalence of the dynamic-deformation-theory wave function with the superposition of multiparticle-multihole wave functions	11
7.	Schematic picture of the rotation of axes and the passage from the lab to the intrinsic frame of reference	15
8.	The Beta-Gamma Mesh Used for Integrations Over β and γ	29
9.	Ideal vibrational spectrum	35
10.	Correspondence Principle for Phonon States of the Vibrational Model and Rotational Bands of the Rotational Model	50
11.	Deformed Single-Particle Levels for A Neutron or A Proton	62
12.	Excitation Energies of Lowest 2+ States (A = 12-240)	89
13.	Excitation Energies of Lowest 4+ States (A = 12-240)	90
14.	Excitation Energies of $0'^+$ (0_β) States (A = 12-240)	91
15.	Excitation Energies of $2'^+$ (2_γ) States (A = 12-240)	92
16.	B(E2; $0^+ \to 2^+$) Values (A = 12-240)	93
17.	Quadrupole Moments of Lowest 2^+ States (A = 12-240)	94
18.	Magnetic Moments of Lowest 2^+ States (A = 12-240)	95
19.	Low-Energy States of ^{16}O	97
20.	Contour Plot of the Potential Energy of ^{16}O	98
21.	Low-Energy States of ^{72}Se	100
22.	Fission Isomers of ^{240}Pu	101
23.	Rotational-Vibrational Bands in ^{168}Er	103
24.	Rotational-Vibrational Bands in ^{164}Er	104
25.	Rotational-Vibrational Bands in ^{166}Er	105

List of Tables

		Page
1.	Symmetries and Quantum Numbers of the Single-Particle Basis and Final Wave Functions of the Dynamic Deformation Theory	12
2.	Estimated Values of I_{max}, N_β	14
3.	Allowed K values and basic tensor components of collective wave functions	31
4.	Wave Functions and Some Expectation Values for the $N \leq 3$ States of the Vibrational Model	41
5.	Signs and Magnitudes of E2 Matrix Elements in the Vibrational Model	45
6.	Signs and Magnitudes of E2 Matrix Elements in the Rotational Model	53
7.	Signs and Magnitudes of Some E2 Matrix Elements Utilized for the "Rotational" Phase Convention	54
8.	Electromagnetic Moments of ^{164,166,168}Er	106
9.	Calculated K-Components and Other Properties of ^{168}Er	107

Preface

The various attempts at unifying the many aspects of nuclear properties can be classified into four types:

  I. Unified theory of spherical-transitional-deformed even-even nuclei.
 II. Unified theory of light-medium-heavy even-even nuclei.
III. Unified theory of even-even-, odd-A, and odd-odd nuclei.
 IV. Unified theory of nuclear structure and reactions.

The first of these unifications has been achieved in several different ways - for instance, via the Bohr-Hamiltoninan-Method, the Quasi-Boson-Method, and via the Interacting-Boson-Method.

The second unification, which includes the first one, has recently been attempted on the basis of the Dynamic Deformation Theory. This first attempt of its kind is the subject of the following lecture notes.

This theory represents a new combination of several old methods and models: Bohr-Hamiltonian-Method, Rainwater-Nilsson model, BCS method, Strutinsky method, and the Cranking method.

After a brief discussion of the historical background, the methods and models mentioned above are discussed. Differences from the earlier versions are pointed out. The various concepts and mathematical techniques of possible general applicability are described in some detail, while the technical details are kept to a minimum.

Some of the results discussed below have been presented at two recent conferences (A = 12-240: Rhodos, May 1979; ^{16}O, ^{72}Se, ^{240}Pu: Nashville, September 1979). Others ($^{164,166,168}Er$) are completely new.

These results are discussed from the point of view of the first two unifications. On the basis of these, suggestions are presented for future studies leading to the third and the fourth unifications.

I take this opportunity to thank the following for their moral and/or financial support during the course of the development of the theory presented here: Professor J.H. Hamilton (Vanderbilt University, Nashville, USA), Dr. R. Foucher (IPN, Orsay, France), Dr. D. Gogny (CEN, Bruyeres-le-chatel, France), Dr. J.S. Vaagen (University of Bergen), Mme J. Delaunay (CEN, Saclay, France), and Professor D.A. Bromley (Yale University, USA).

I particularly thank Professor Aage Bohr (NBI, Copenhagen, Denmark) for his inspiration and encouragement throughout the development of this work.

- Krishna Kumar
Bergen, 1979

"We dance around in a ring and suppose,
But the secret sits in the middle and knows."

-Carl Sandburg
Poetry, April 1936

"In a way, science might be described as paranoid thinking applied to Nature: we are looking for natural conspiracies, for connections among apparently disparate data... The search for patterns without critical analysis, and rigid skepticism without a search for patterns, are the antipodes of incomplete science. The effective pursuit of knowledge requires both functions."

-Carl Sagan
The Dragons of Eden, 1977

I. Introduction

A. Historical Background

We start with a brief discussion of the historical background relevant to the present subject. The following list is hardly exhaustive. A more complete list can be found in the textbooks (for instance, Brown 1967, de-Shalit & Feshbach 1974, Preston & Bhaduri 1975, Bohr & Mottelsen 1975).

Casimir (1936) pointed out that the "large" values of the nuclear quadrupole moments, deduced from the hyperfine splittings of atomic levels, can be understood in terms of a deformed (spheroidal) nuclear shape (see fig. 1).

Mayer (1949) and Haxel et al. (1949) showed that many nuclear properties can be understood in terms of a nuclear shell model — where a nucleon moves almost freely in the average field generated by the rest of the nucleons (see fig. 2).

Mayer (1950) showed that the success of the simple, single-particle shell model can be related to the short range of the nucleon-nucleon interaction which lowers the level with $I = j$ compared to other I values allowed by the angular momentum coupling of the valence nucleons (odd number) in the j-orbit.

In the same year, two other important papers were published. Racah (1950) proposed a new classification of nuclear states based on a seniority quantum number which counts the number of unpaired nucleons. Rainwater (1950) extended Casimir's idea of nuclear deformation by arguing that if the nucleus is longer in one direction than the other two ($R_z > R_x = R_y$), then the corresponding frequencies of the average field must be split according to
$$\omega_z < \omega_x = \omega_y.$$
This simple argument allowed Rainwater to generalize the spherical-single-particle model of Mayer-Jensen to a deformed-single-particle model.

Bohr (1952) proposed a collective Hamiltonian and a collective Schrödinger equation for the problem of five-dimensional harmonic-quadrupole vibrations of the nucleus as a whole:

$$H_{coll} = \tfrac{1}{2} C \sum_\mu |\beta_\mu|^2 + \tfrac{1}{2} B \sum_\mu |\dot\beta_\mu|^2 \qquad (1.1)$$

$$= \tfrac{1}{2} C \beta^2 + \tfrac{1}{2} B (\dot\beta^2 + \beta^2 \dot\gamma^2) + \tfrac{1}{2} \sum_{k=1}^{3} \mathcal{J}_k \omega_k^2. \qquad (1.2)$$

This important paper was neglected for many years, perhaps because the problem of harmonic vibrations can be solved more easily in terms of phonons or bosons.

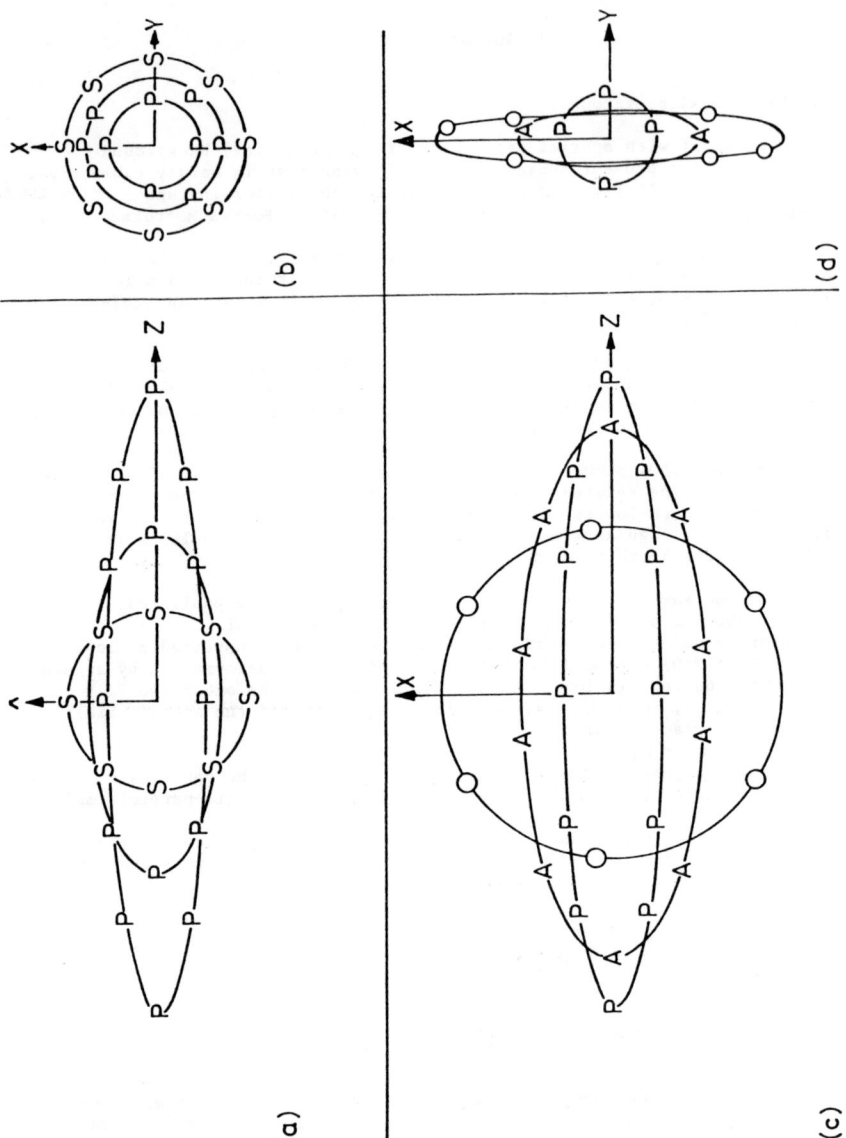

Fig. 1. Projections of nuclear shapes on the z-x, x-y planes. The Hill-Wheeler definition of nuclear deformations has been employed (see sec. IV.A).
1.a,b: $\gamma = 0^0$, $\beta = 0.0$ (S), 0.4 (P), 0.8 (P). 1.c,d: $\beta = 0.8$, $\gamma = 0^0$ (P), 30^0 (A), 60^0 (O).

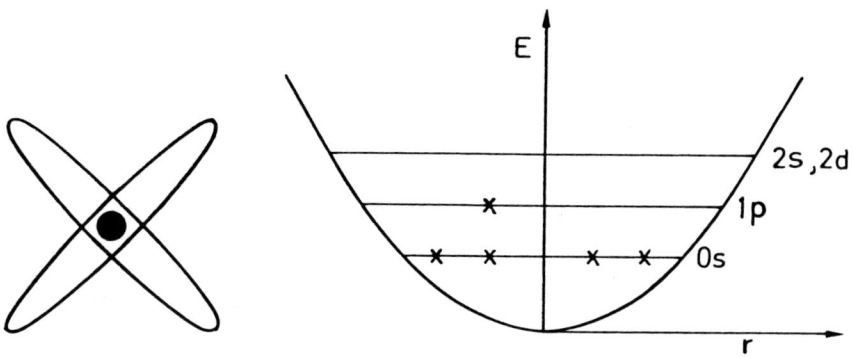

(a) NUCLEAR SHELL MODEL   (b) OSCILLATOR MODEL

Fig. 2. a: Schematic picture of the nuclear shell model. In contrast to the atomic shell model, the nucleonic orbits remain close to the "core" and interact with it more strongly.

2. b: Lowest few single-particle levels of the harmonic oscillator model.

Hill and Wheeler (1953) emphasized the importance of the γ degree of freedom. They pointed out that a deformed nucleus can go from a deformed, prolate (P) shape to a deformed, oblate (O) shape via the γ degree of freedom instead of having to go through a barrier at the spherical (S) shape (see fig. 3).

In the same paper, Hill and Wheeler pointed out that when one introduces the collective degrees of freedom (for instance, deformation β), one is introducing extra degrees of freedom. In order to overcome this problem, they proposed the generator-coordinate-method where the deformed wave function $\phi(\vec{r},\beta)$ is replaced by

$$\Psi(\vec{r}) = \int \phi(\vec{r},\beta) \, f(\beta) \, d\beta. \qquad (1.3)$$

In the same year Bohr and Mottelsen (1953) presented the nuclear rotational model where the nucleus with a permanent deformation ($\beta > 0$, $\gamma = 0°$: axial symmetry) rotates around an axis perpendicular to the symmetry axis (see Fig. 4). This important model provides the basis of many of the most successful methods of analyzing nuclear spectra, and nuclear reactions.

Inglis (1955) proposed the Cranking method of calculating the moment of inertia associated with nuclear rotations. In the same year, Nilsson (1955) presented an extensive calculation of the deformed single-particle (Nilsson) levels.

Bardeen, Cooper and Schrieffer (1957) presented the BCS method for pairing correlations in superconductors.

Bohr, Mottelsen, and Pines (1958) showed that the BCS type pair correlations play a crucial role in atomic nuclei. In the same year, Bogolyubov (1958) provided the theoretical foundation for the BCS theory via the Bogolyubov transformation from pure particles, holes to their linear combinations called Quasi-particles:

$$a^+ = u \, c^+ + v \, c.$$

In the same year, Elliot (1958) proposed a quadrupole-quadrupole nuclear force which leads to nuclear deformations and rotational spectra.

Kisslinger and Sorensen (1960) presented the first detailed calculations of the collective spectra of spherical nuclei starting from the pairing-plus-quadrupole model. They employed the Random Phase Approximation (RPA) or the Quasi-Boson method.

Baranger (1963) presented the Hartree-Bogolyubov treatment of the pairing-plus-quadrupole model. In the same year, Kumar (1963) presented a numerical method of solving Bohr's collective Hamiltonian which allowed for the treatment of spherical, transitional, deformed even-even nuclei with the same method.

Strutinsky (1966) presented a shell correction method which was later shown to yield the Hartree-Fock type of accuracy with much smaller computation times.

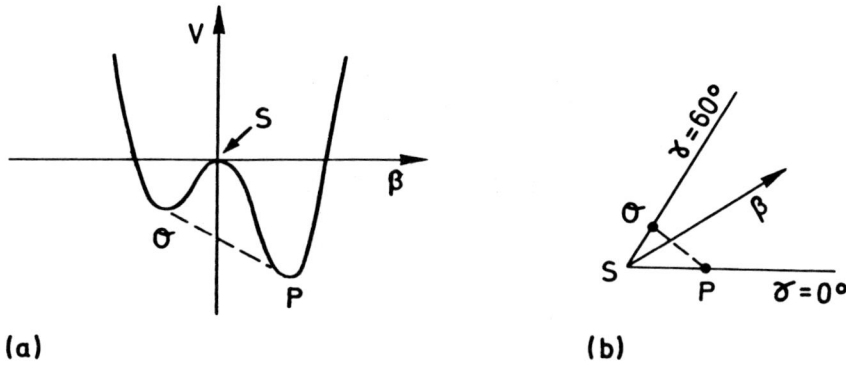

Fig. 3. a: Potential energy vs. deformation for a deformed, prolate nucleus.
3. b: The passage from prolate to oblate shapes along the γ degree of freedom.

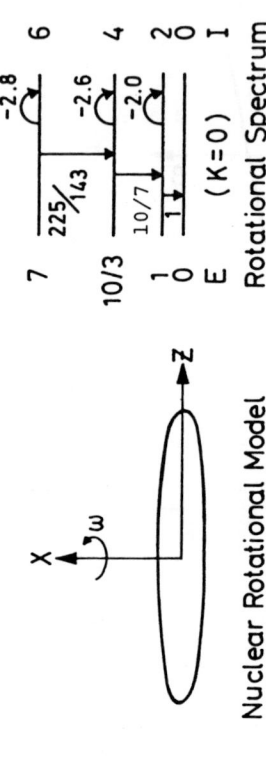

Fig. 4. a: Schematic picture of the nuclear rotational model.
4. b: Ideal rotational spectrum with the associated B(E2) values (↓) and spectroscopic quadrupole moments (↷).

Bohr and Mottelsen (1969) presented a very general discussion of
the symmetries relevant to physical phenomena. Myers and Swiatecki (1974)
generalized the liquid drop model to the droplet model. Bohr and Mottelsen
(1975) discussed many concepts and models relevant to the theory of nuclear
structure. Yariv, Ledergerber, and Pauli (1976) presented two very useful
stationarity conditions for the shell correction energy of the Strutinsky
Method.

Kumar, Remaud, Aguer, Vaagen, Rester, Foucher, and Hamilton (1977)
combined the ideas of many of the papers mentioned above and developed a
scaling method which allowed for the possibility of quick microscopic
calculations in a large configuration space. Thus, it became possible to
(a) discard the desegregration of nucleons into an "inert" group and a
"valence" group, (b) to remove fitting parameters like effective charges
and inertial renormalization factors, and (c) to extend the first unification to light-medium-heavy nuclei.

I.B. Different Methods of Constructing a Many-Body Wave Function.

1. Particle-Hole Method

This is a very basic method. A nice discussion has been given by
Brown (1967). It is illustrated in Fig. 5 in obvious notation.

2. j^N Method

The total wave function for angular momentum I, projection M is
written as

$$\Psi_{IM} = \sum_{n_1 n_2 \ldots n_\Omega} C_{n_1 n_2 \ldots n_\Omega} \left(j_1^{n_1} j_2^{n_2} \ldots j_\Omega^{n_\Omega} \right)$$

This method has been exploited quite extensively in the conventional
shell model calculations, see for instance, McGrory and Wildenthal (1980).

3. Quasi-Boson (RPA) Method

A boson operator of a given angular momentum J and Projection M_J
is constructed out of pairs of nucleons:

$$A^+_{JM_J} = \sum_{j_1 j_2 m_1 m_2} \left(C^+_{j_1 m_1} C^+_{j_2 m_2} \right)_{JM_J}.$$

The total wave function is written as

$$|IM> = \sum_n C_n \left[A^+_J \otimes A^+_J \otimes \ldots A^+_J \right]_{IM} |->,$$

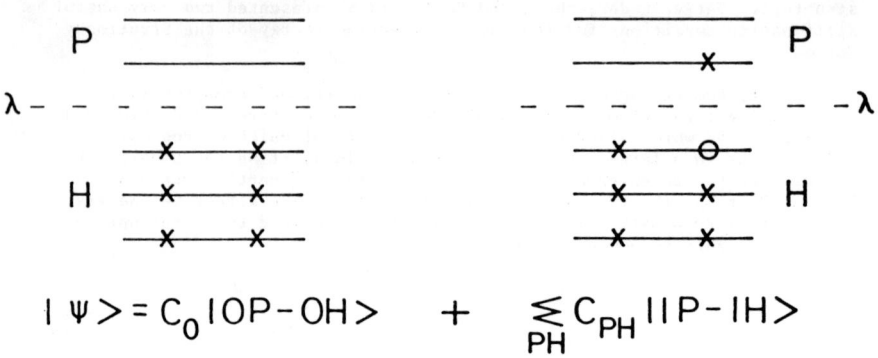

Fig. 5. Schematic picture of the particle-hole method of constructing the many-body nuclear wave function.

where n is the number of bosons coupled to I,M; and $|->$ is a vacuum with respect to the boson operators. These operators are generally not true boson operators. Hence, the method is often called the quasi-boson method or the RPA method, implying that the non-boson parts of the commutation relations are assumed to cancel because of random phases [see, for instance, Kisslinger & Sorensen (1960), Sørensen (1970), Kishimoto and Tamura (1976)].

4. Generator-Coordinate-Method

This method, proposed by Hill and Wheeler (1953), has been discussed above. The generating function, $f(\beta)$, of eq. (1.3) is determined via a minimization of $(\Psi^*H\Psi)/(\Psi^*\Psi)$ with respect to variations in $f(\beta)$. A gaussian overlap approximation is often employed which leads to RPA type of equations valid in the limit of small amplitude (or nearly harmonic) vibrations. One could avoid this approximation (at least in principle) by employing a discrete mesh in β and calculating the matrix H_{ij}, where i and j refer to different mesh points. But the calculation of overlaps of wave functions of different deformations is a time-consuming procedure and the computation time increases rapidly with the size of the configuration space and with the number of collective degrees of freedom.

5. Dynamic Deformation (Collective Hamiltonian) Method

A formal derivation of the collective Hamiltonian of type (1.1-1.2), starting from the generator-coordinate-method, has been attempted by Giraud and Grammaticos (1974). However, their procedure leads in some cases to negative masses. A simple argument, but not a formal derivation, is given below.

Starting with the wave function of eq. (1.3), the energy expectation value is written as (assuming Ψ to be normalized)

$$E = \int d\beta' f(\beta') \int d\vec{r}' \Phi(\vec{r}',\beta') H(\vec{r}',\vec{r}) \Phi(\vec{r},\beta) d\vec{r} f(\beta) d\beta$$

$$= \int d\beta' f(\beta') H_{coll}(\beta',\beta) f(\beta) d\beta$$

$$= \sum_{ij} f_j \left(H_{coll}\right)_{ji} f_i ,$$

where $H_{coll} = V(\beta) + T_{coll}\left(\beta, \frac{\partial}{\partial\beta}\right)$,

$\left(H_{coll}\right)_{ij} = <\beta_i |H_{coll}|\beta_i> = V(\beta_i)$,

and $\left(H_{coll}\right)_{ij} = <\beta_j |H_{coll}|\beta_i> = <\beta_j|T_{coll}|\beta_i>$ $(j \neq i)$.

Thus, the potential energy of the collective Hamiltonian is identified with the local $(\beta' = \beta)$ part of the GCM energy function, while the kinetic energy is identified with the non-local part $(\beta' \neq \beta)$ of the same function. This is a conceptual connection only. The expressions used in practical calculations to be discussed below are based on the cranking method (see sec. II).

The dynamic deformation theory wave functions, obtained as solutions of the collective Hamiltonian, may be pictured as linear combinations of microscopic wave functions corresponding to different shapes (see Fig. 6). As deformation changes from β_1 to β_2 to β_3, the single-particle levels move up and down. Thus, although we may choose the simplest nucleon distributions for each deformation (for instance, OP-OH for e-e nuclei without pairing or zero-quasi-particle states with pairing), our final wave function represents a linear combination of quite complicated nucleon distributions. Thus, the dynamic deformation theory allows us to include in a simple way multi-particle-multi-hole (or multi-quasi-particle) states although we work with only OP-OH (or zero-quasi-particle) states at each intermediate step. To be sure, all possible states are not included. The included states are determined by the assumed symmetries. Main approximations of the DDT are discussed below.

I. C. Main Approximations of the Dynamic Deformation Theory

There are two main approximations.

1. States of a certain symmetry separate from all other states. There are the fundamental symmetries: parity and rotational invariance which lead to the good quantum numbers Π, I, M. These and the additional symmetries exploited in the DDT are listed in Table 1.

Note that although I,M (or j,m) are not good quantum numbers for the deformed single-particle basis, they are good for the final wave functions.

Furthermore, although only the zero-quasi-particle states are employed explicitly for parts of the calculation, effects of two-quasi-particle states are included via the collective kinetic energy terms and effects of 4,8,... quasi-particle states are included via the dynamics (see sect. I.B.).

2. The second main approximation of the DDT is the adiabatic approximation, that the collective frequencies are small compared to single-particle frequencies. In order to <u>estimate</u> the extreme limits of the validity of this approximation, we employ the rotational model of Bohr and Mottelsen. Then, the adiabatic condition for rotations gives

$$\hbar\omega_R \ll \hbar\omega_{sp}. \qquad (1.4)$$

Or, $\quad \dfrac{\hbar^2 I}{\mathcal{J}} = \dfrac{IE_2}{3} \ll \hbar\omega_{sp}.$

Or, $\quad I_{max} = \dfrac{3\hbar\omega_{sp}}{E_2}. \qquad (1.5)$

Similarity, the adiabatic condition for β-vibrations gives

$$\hbar\omega_\beta \ll \hbar\omega_{sp}. \qquad (1.6)$$

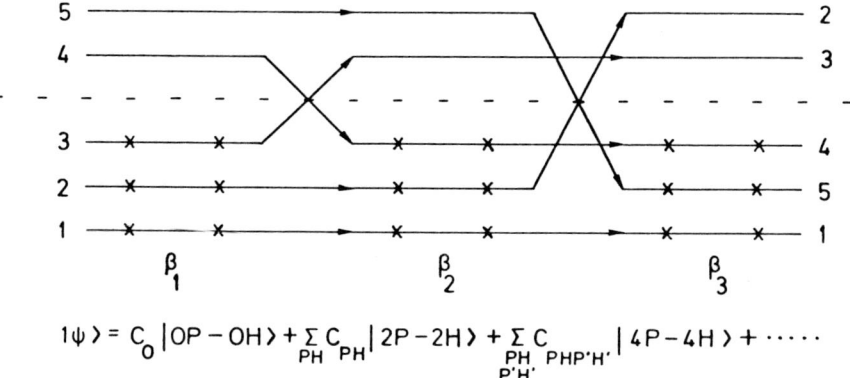

$|\psi\rangle = C_0 |OP-OH\rangle + \sum_{PH} C_{PH} |2P-2H\rangle + \sum_{\substack{PH \\ P'H'}} C_{PHP'H'} |4P-4H\rangle + \cdots$

Fig. 6. Schematic picture of the equivalence of the dynamic-deformation-theory wave function with superposition of multiparticle-multihole wave functions.

Table 1. Symmetries and Quantum Numbers of the Single-Particle-Basis and Final Wave Functions of the Dynamic Deformation Theory. The relevant symmetry (quantum number) is indicated by a cross. The final wave functions refer to the low-energy states of even-even nuclei.

Symmetry (Quantum Number)	S. P. Basis		Final Wave Function
	Spherical	Deformed	
Parity (Π)	x	x	x
Radial (N)	x		
Orbital (L)	x		
Rotational: I	x		x
M	x		x
Isopin symmetry [a]	x	x	
Time-Reversal : Symmetry [b]			x
: Degeneracy of Time-Reversed States	x	x	x
Seniority (Number of unpaired nucleons) [b]			x
Rotation via R_k (π) [Rotation of axes by π keeping axis k fixed] [c]			x

[a] This occurs in the present version of DDT because of the scaling method of Kumar et al. (1977).

[b] The microscopic wave functions correspond to BCS type zero-quasi-particle states. Hence, they are symmetric under time-reversal and their seniority quantum number is zero.

[c] The quantum numbers $r_1 = r_2 = r_3 = +1$ are employed to restrict the K values to $0, 2, \ldots, I$ (I = even) or $2, 4, \ldots, I-1$ (I = odd), and to choose the phase factor connecting these with their $K < 0$ counterparts.

Identifying $\hbar\omega_\beta$ with the excitation energy of the first excited 0^+ state, we define

$$n_\beta = \frac{\hbar\omega_{sp}}{\hbar\omega_\beta} = \frac{\hbar\omega_{sp}}{E_0'} \quad . \tag{1.7}$$

Using the relation

$$\hbar\omega_{sp} = 41 \, A^{-1/3} \text{ MeV} , \tag{1.8}$$

and the observed values of E_2, E_0' we calculate I_{max}, n_β for a few typical nuclei. The corresponding values are listed in Table 2. The ratios $1/I_{max}$, $1/n_\beta$ provide estimates of the smallness of the ratio of collective to s.p. frequencies.

As expected, the adiabatic approximation gets worse in the spherical limit (^{204}Pb in the table) and in lighter nuclei. But the estimates given in Table 2 show that the approximation is valid for the low-spin states of practically all even-even nuclei.

II. Microscopic Theory of Collective Motion

(How to go from the Time-Dependent Schrödinger Equation to the Collective Schrödinger Equation?)

The present derivation is based on the cranking method of Inglis (1955).

Consider the time-dependent Schrödinger equation

$$H\psi = i\hbar\frac{\partial\psi}{\partial t} \quad . \tag{2.1}$$

Now, consider the same equation in a rotating frame, a frame attached to the nucleus rotating with angular velocity ω around the z-axis (see Fig. 7):

$$H'\psi' = i\hbar\frac{\partial\psi'}{\partial t} . \tag{2.2}$$

According to quantum mechanics (e.g. Messiah 1962), the wave functions ψ', ψ are related by the equation

$$\psi' = R_\omega \psi , \tag{2.3}$$

where $R_\omega = e^{i\phi L_z}$. $\tag{2.4}$

On combining eqs. (2.2-2.4), we get

$$H'R_\omega\psi = i\hbar\left(\left[\frac{\partial}{\partial t} R_\omega\right]\psi + R_\omega\frac{\partial\psi}{\partial t}\right)$$

$$= i\hbar\left(i\omega L_z R_\omega\psi + R_\omega\frac{\partial\psi}{\partial t}\right)$$

$$= -\hbar\omega L_z R_\omega\psi + i\hbar R_\omega\frac{\partial\psi}{\partial t} ,$$

where we have employed $\omega = \frac{\partial\phi}{\partial t}$. On rearranging terms on the two sides of the equation, we get

Table 2. Estimated Values of I_{max}, n_β.
The ratios I/I_{max} and $1/n_\beta$ represent ratios of $\omega_{coll}/\omega_{sp}$ for rotations and vibrations, respectively.

Nucleus	E_2 (Mev)	$E_{0'}$ (Mev)	$\hbar\omega_{sp}$ (Mev)	I_{max}	n_β
^{12}C	4.439	7.655	17.9	12	2.3
^{16}O	6.919	6.050	16.3	7	2.7
^{168}Er	0.080	1.217	7.4	279	6.1
^{204}Pb	0.899	1.585	7.0	23	4.4
^{240}Pu	0.043	0.861	6.6	460	7.7

- 15 -

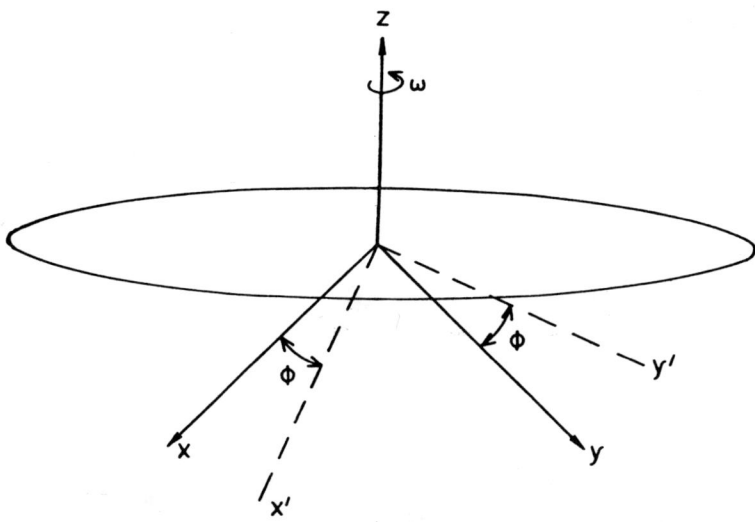

Fig. 7. Schematic picture of the rotation of axes and the passage from the lab to the intrinsic frame of reference.

$$(H' + \hbar\omega L_z) R_\omega \psi = i\hbar R_\omega \frac{\partial \psi}{\partial t} \ . \qquad (2.5)$$

Operate on both sides of eq. (2.1) by R_ω:

$$R_\omega H \psi = i\hbar R_\omega \frac{\partial \psi}{\partial t} \ .$$

Or, $$HR_\omega \psi = i\hbar R_\omega \frac{\partial \psi}{\partial t} \ , \qquad (2.6)$$

where we have used the rotational invariance of H, i.e.

$$[H, R_\omega] = HR_\omega - R_\omega H = 0 \ . \qquad (2.7)$$

Comparison of Eqs. (2.5, 2.6) gives

$$H' + \hbar\omega L_z = H$$

Or, $$H' = H - \hbar\omega L_z \ . \qquad (2.8)$$

Rewrite eq. 2.8 (using $\omega = \dot{\phi}$, $L_z = \frac{1}{i}\frac{\partial}{\partial \phi}$) as

$$H' = H - \frac{\hbar}{i} \dot{\phi} \frac{\partial}{\partial \phi} \ . \qquad (2.9)$$

For a set of collective variables $\dot{\alpha}_\mu$, we generalize eq. (2.9) to

$$H' = H - \frac{\hbar}{i} \sum_\mu \dot{\alpha}_\mu \frac{\partial}{\partial \alpha_\mu} \ . \qquad (2.10)$$

Now we treat the $\dot{\alpha}_\mu$ term of eq. (2.10) as a small time-dependent perturbation, that is we invoke the adiabatic approximation. The basis is defined by the eigenvalue equation

$$H'|n\rangle = H|n\rangle = E_n|n\rangle \quad (H' = H \text{ for } \dot{\alpha}_\mu = 0), \quad (2.11)$$

where H is the microscopic many-body Hamiltonian (e.g. the pairing-plus-quadrupole or the Nilsson-plus-pairing Hamiltonian) and n is an index for the nuclear many-body states:

$$n = \begin{cases} 0P-0H, \ 1P-1H, \ 2P-2H, \ldots \\ \text{Or } 0QP, \ 2QP, \ 4QP, \ldots \end{cases} \text{e-e}$$

$$= \begin{cases} 1P, \ 2P-1H, \ 3P-2H, \ldots \\ 1QP, \ 3QP, \ 5QP, \ldots \end{cases} \text{odd-A} \qquad (2.12)$$

$$= \begin{cases} 2P-0H, \ 3P-1H, \ 4P-2H, \ldots \\ 2QP, \ 4QP, \ 6QP, \ldots \end{cases} \text{odd-odd} \ .$$

For e-e nuclei, one gets

$$E'_n = E_n - \hbar^2 \sum_{\mu\nu \atop m} \dot{\alpha}_\mu \dot{\alpha}_\nu \frac{|\langle n|\frac{\partial}{\partial \alpha_\mu}|m\rangle||\langle m|\frac{\partial}{\partial \alpha_\nu}|n\rangle|}{E_m - E_n}$$

$$= E_n - \frac{1}{2} \sum_{\mu\nu} \dot{\alpha}_\mu \dot{\alpha}_\nu B_{\mu\nu}(\alpha) , \qquad (2.13)$$

where $B_{\mu\nu}(\alpha) = 2\hbar^2 \sum_{nm} \frac{|<n|\frac{\partial}{\partial \alpha_\mu}|m>||<m|\frac{\partial}{\partial \alpha_\nu}|n>|}{E_m - E_n}$. (2.14)

Now we have to keep in mind that the true Hamiltonian is H, not H'. We need to think of the $\dot{\alpha}_\mu$ terms, not as true corrections to the Hamiltonian, but as constraining terms. The constraints are given by

$$\sum_\nu B_{\mu\nu}(\alpha) \dot{\alpha}_\nu = \frac{\hbar}{i} <n'|\frac{\partial}{\partial \alpha_\mu}|n'> . \qquad (2.15)$$

They are satisfied automatically in the perturbation theory. The constrained energy is given by

$$H_{coll} = <n'|H|n'>$$
$$= <n'|H' + \frac{\hbar}{i} \sum_\mu \dot{\alpha}_\mu \frac{\partial}{\partial \alpha_\mu}|n'>$$
$$= E_{n'} + \sum_\mu \dot{\alpha}_\mu \frac{\hbar}{i} <n'|\frac{\partial}{\partial \alpha_\mu}|n'>$$
$$= E_{n'} + \sum_{\mu\nu} \dot{\alpha}_\mu \dot{\alpha}_\nu B_{\mu\nu}(\alpha)$$
$$= E_n + \frac{1}{2} \sum_{\mu\nu} \dot{\alpha}_\mu \dot{\alpha}_\nu B_{\mu\nu}(\alpha) .$$

Or, $H_{coll} = V(\alpha) + \frac{1}{2} \sum_{\mu\nu} \dot{\alpha}_\mu \dot{\alpha}_\nu B_{\mu\nu}(\alpha)$, (2.16)

where $V(\alpha) \equiv E_n = <n(\alpha)|H|n(\alpha)>$, (2.17)
is the potential energy of deformation and the mass-parameter-matrix $B_{\mu\nu}(\alpha)$ is given by eq. (2.14).

Usually n indicates OP-OH or OQP states of e-e nuclei and the index n is dropped.

The Hamiltonian of eq. (2.16) is a classical Hamiltonian. It is quantized by employing the Pauli method (following Bohr 1952),

$$\hat{H}_c\left(\alpha_\mu, \frac{\partial}{\partial \alpha_\mu}\right) = V(\alpha_\mu) - \frac{\hbar^2}{2} D^{-\frac{1}{2}} \sum_{\mu\nu} \frac{\partial}{\partial \alpha_\mu} D^{\frac{1}{2}} B_{\mu\nu}^{-1} \frac{\partial}{\partial \alpha_\nu} , \qquad (2.18)$$

where D is the determinant and B^{-1} the inverse of the matrix $B_{\mu\nu}$. The corresponding collective Schrödinger equation is

$$\hat{H}_c \Psi_{IM}(\alpha) = E_I \Psi_{IM}(\alpha) . \qquad (2.19)$$

The equations (2.18, 2.19) are valid for any collective variables. For the case of small amplitude, quadrupole vibrations around the spherical shape, they were first given by Bohr (1952):

$$\alpha_\mu \equiv \alpha_{2\mu} \quad (\mu = 0, \pm 1, \pm 2), \tag{2.20}$$

$$V(\alpha) = \frac{1}{2} C \sum_\mu \alpha_\mu^2, \tag{2.21}$$

$$B_{\mu\nu}(\alpha) = B \, \delta_{\mu\nu}. \tag{2.22}$$

These will be reviewed in the next chapter.

Here we note that the above microscopic theory of collective motion is quite general. One can start from any microscopic Hamiltonian H, calculate the potential energy function $V(\alpha)$ and the mass-parameter-matrix $B_{\mu\nu}(\alpha)$, and solve the CSE (2.19) to obtain the energies and wave functions of the collective states.

Although the derivations given in the literature are quite different (and much more lengthy), the physics leading to the collective Hamilttonian is basically the same. Thus, the collective Hamiltonian for five-dimensional quadrupole vibrations has been employed for spherical-transitional-deformed even-even nuclei by Kumar (1963, 1974, 1975a, 1977a), Kumar & Baranger (1968), and by Dobaczewski et al. (1975). A similar type of Hamiltonian has been employed by Bes et al. (1970) for the problem of pair-vibrations.

The DDT discussed below includes nine-dimensional collective motion: five-dimensional quadrupole motion (3-D rotations + β-vibrations + γ-vibrations) plus four-dimensional pairing motion (proton energy gap, proton Fermi energy, neutron energy gap, neutron Fermi energy).

III. Bohr's Collective Hamiltonian

A. The Hamiltonian and its Symmetries

In this chapter, we discuss the symmetries and properties of Bohr's collective Hamiltonian for the problem of five-dimensional quadrupole vibrations. In the lab-frame of axes, this Hamiltonian is given by eq. (2.16) with

$$\alpha_\mu \equiv \alpha_{2\mu} \quad (\mu = -2, -1, 0, 1, 2). \tag{3.1}$$

It is written in the intrinsic frame by employing the transformation

$$\alpha_\mu = \sum_a \mathcal{D}^2_{\mu a}(\phi,\theta,\psi)\, \beta'_a, \tag{3.2}$$

where ϕ, θ, ψ are the Euler's angles connecting the two sets of axes. The intrinsic axes are chosen such that

$$\beta'_1 = \beta'_{-1} = 0, \quad \beta'_2 = \beta'_{-2}. \tag{3.3}$$

Following Bohr (1952), we define

$$\beta'_2 = \beta'_{-2} = \frac{1}{\sqrt{2}} \beta \sin\gamma, \quad \beta'_0 = \beta \cos\gamma. \tag{3.4}$$

Following Kumar (1963), we define

$$\beta_0 = \beta'_0 = \beta \cos\gamma, \quad \beta_{2'} = \sqrt{2}\,\beta'_2 = \sqrt{2}\,\beta'_{-2} = \beta \sin\gamma. \tag{3.5}$$

With these transformations (and utilizing the properties of the $\mathcal{D}$-functions), the collective Hamiltonian for 5-D quadrupole motion can be rewritten as

$$H_{coll} = V(\beta_0, \beta_{2'}) + \frac{1}{2} \sum_{ab=0,2'} B_{ab}(\beta_0,\beta_{2'}) \dot{\beta}_a \dot{\beta}_b \tag{3.6}$$
$$+ \frac{1}{2} \sum_{k=1,2,3} \mathcal{J}_k(\beta_0,\beta_{2'}) \omega_k^2,$$

where $\mathcal{J}_k$ is the moment of inertia for rotation around the intrinsic-k-axis and ω_k is the angular frequency.

In terms of the 'polar' coordinates (β,γ), the same Hamiltonian takes the form

$$H_{coll} = V(\beta,\gamma) + \frac{1}{2} B_{\beta\beta}(\beta,\gamma)\dot{\beta}^2 + B_{\beta\gamma}(\beta,\gamma)\dot{\beta}\dot{\gamma}$$
$$+ \frac{1}{2}B_{\gamma\gamma}(\beta,\gamma)\beta^2\dot{\gamma}^2 + \frac{1}{2}\sum_{k=1,2,3}\mathcal{J}_k(\beta,\gamma)\omega_k^2 \quad . \tag{3.7}$$

This Hamiltonian depends on seven functions of deformation: V, $B_{\beta\beta}$, $B_{\beta\gamma}$, $B_{\gamma\gamma}$, $\mathcal{J}_1$, $\mathcal{J}_2$, and $\mathcal{J}_3$. These functions are capable of describing an ideal vibrational (spherical) nucleus (see sect III.C), an ideal rotational (deformed) nucleus (see sect. III.D), as well as realistic nuclei with various degrees of anharmonicity or rotation-vibration coupling. Before discussing the respective solutions, we give a brief review of the symmetry properties of these functions.

The Hamiltonian must be a scalar. Thus, it can be written in terms of the basic scalars formed by the quadrupole tensors β_a, $\dot{\beta}_a$. One has to keep in mind that while $\beta_1 = \beta_{-1} = 0$, $\dot{\beta}_1$, $\dot{\beta}_{-1}$ are in general non-zero. Hence, one has to start with the lab-frame form of the Hamiltonian and utilize the transformation to the intrinsic frame only at the last step.

The basic scalars formed from the product of $\alpha_{2\mu}$ tensors are:

$$\alpha_2 \cdot \alpha_2 = \beta^2 , \tag{3.8}$$

$$(\alpha_2 \times \alpha_2)_2 \cdot \alpha_2 \propto \beta^3 \cos 3\gamma \quad . \tag{3.9}$$

All scalar products of α_2 tensors can be written as simple products or sums of the two basic scalars of eqs. (3.8-9). Therefore, the potential energy must have the form

$$V = V(\beta^2, \beta^3 \cos 3\gamma). \tag{3.10}$$

In particular, there should be no linear term in β in the potential function. In other words, we must have

$$\partial V/\partial \beta = 0 \text{ at } \beta = 0. \tag{3.11}$$

Thus, the potential function must have a minimum or a maximum at $\beta = 0$. A parabolic expansion around $\beta = 0$ is fine. But a parabolic expansion around $\beta \neq 0$ is not correct.

As regards the six inertial functions, their β-γ dependence is much more complicated because of the time-dependence of the quadrupole tensor. In the spherical limit ($\beta \to 0$), the β-γ-dependence of the six inertial functions is given by (Bohr 1952)

$$B_{\beta\beta} = B_{\gamma\gamma} = B, \quad B_{\beta\gamma} = 0, \tag{3.12}$$

$$\mathcal{J}_k = 4B\beta^2 \sin^2\left(\gamma - \frac{2}{3}k\pi\right), \tag{3.12b}$$

where B is a constant (compare Eqs. 1.2 and 3.7). The most general case has been discussed by Kumar and Baranger (1967). Their results are given below.

The six inertial functions are written in terms of six V-type functions (i.e. functions whose β-γ-dependence comes only from sums and products of β^2 and $\beta^3 \cos 3\gamma$) denoted by: b_1, b_2, b_3, b_4, b_5, b_6. The three moments of inertia are given by (k = 1,2,3)

$$\mathcal{J}_k(\beta,\gamma) = 4\, B_k(\beta,\gamma)\beta^2 \sin^2\left(\gamma - \tfrac{2}{3}k\pi\right), \qquad (3.13)$$

$$\begin{aligned}
B_3(\beta,\gamma) &= b_1 - 14\left(\beta^2 b_4 - b_5\beta^3 \cos 3\gamma + \beta^4 b_6\right) \\
&\quad + \beta\cos\gamma\left[b_2 - 20\left(\beta^2 b_5 - 2 b_6 \beta^3 \cos 3\gamma\right)\right] \\
&\quad + \beta^2 \cos 2\gamma\left[-b_3 + 20\left(b_4 - \beta^2 b_6\right)\right],
\end{aligned} \qquad (3.14)$$

$$B_1(\beta,\gamma) = B_3(\beta, \gamma - 120^\circ), \qquad (3.15)$$

$$B_2(\beta,\gamma) = B_3(\beta, \gamma + 120^\circ). \qquad (3.16)$$

The three vibrational mass parameters are given by

$$\begin{aligned}
B_{\beta\beta}(\beta,\gamma) &= b_1 + \beta^2\left[b_3 + 6\left(6 b_4 + \beta^2 b_6\right)\right] \\
&\quad - \beta\cos 3\gamma\left[b_2 + 6\left(6\beta^2 b_5 - 5 b_6 \beta^3 \cos 3\gamma\right)\right],
\end{aligned} \qquad (3.17)$$

$$B_{\beta\gamma}(\beta,\gamma) = \beta\sin 3\gamma\left[b_2 + 15\left(\beta^2 b_5 - 2 b_6 \beta^3 \cos 3\gamma\right)\right], \qquad (3.18)$$

and

$$\begin{aligned}
B_{\gamma\gamma}(\beta,\gamma) &= b_1 + \beta^2\left[-b_3 + 6\left(b_4 + 6\beta^2 b_6\right)\right] \\
&\quad - \beta\cos 3\gamma\left[-b_2 + 6\left(\beta^2 b_5 + 5 b_6 \beta^3 \cos 3\gamma\right)\right].
\end{aligned} \qquad (3.19)$$

Note that the relations (3.17-19) are much simpler than those given by Kumar and Baranger (1967). Their relations were given in terms of $B_{00'}$, $B_{02'}$, $B_{2'2'}$. The two sets of mass parameters are related by

$$B_{\beta\beta}(\beta,\gamma) = \tfrac{1}{2}\left[(B_{00} + B_{2'2'}) + \cos 2\gamma(B_{00} - B_{2'2'})\right] + \sin 2\gamma\, B_{02'}, \qquad (3.20)$$

$$B_{\beta\gamma}(\beta,\gamma) = \cos 2\gamma\, B_{02'} - \tfrac{1}{2}\sin 2\gamma(B_{00} - B_{2'2'}), \qquad (3.21)$$

and

$$B_{\gamma\gamma}(\beta,\gamma) = \tfrac{1}{2}\left[(B_{00} + B_{2'2'}) - \cos 2\gamma(B_{00} - B_{2'2'})\right] - \sin 2\gamma\, B_{02'}. \qquad (3.22)$$

The above relations are not employed in the actual calculations of the DDT since the functions $b_1 - b_6$ are, in general, quite complicated functions of β^2 and $\beta^3 \cos 3\gamma$. Instead, the seven functions of deformation are computed numerically, starting from a microscopic theory. They satisfy the symmetry conditions automatically, provided the microscopic theory obeys the appropriate symmetry conditions.

The functional dependence of the seven functions is given above because of two reasons. Firstly, one is sometimes interested to know how the mass parameters look as functions of β and γ. The relations given above provide the necessary restrictions on the allowed functional dependence. Secondly, the above relations can be utilized to give explicilty the symmetry properties of the collective Hamiltonian as follows:

$$H_{coll}(\beta,\gamma) = H_{coll}(\beta,\gamma - 120°) = H_{coll}(\beta,\gamma + 120°)$$
$$= H_{coll}(\beta,-\gamma) = H_{coll}(\beta, 60° - \gamma), \qquad (3.23)$$

$$H_{coll}(\beta, 60° + \gamma) = H_{coll}(\beta, 60° - \gamma). \qquad (3.24)$$

These equations lead to the rule that it is sufficient to vary γ from $0°$ to $60°$ for $\beta > 0$. Once the seven functions are known in the pie shaped region of the $\beta-\gamma$ plane (see Fig. 3b), the symmetry relations (3.23) give them over the entire plane. Furthermore, these symmetry relations, combined with the Hamiltonian of eq. (3.7), give the following rules which are useful for (a) making contour plots of the seven funtions, and (b) for making the allowed expansions in $\beta-\gamma$.

Along the prolate ($\gamma = 0°$) axis, the rules are:

$$\frac{\partial V}{\partial \gamma} = 0 = \frac{\partial B_{\beta\beta}}{\partial \gamma} = \frac{\partial B_{\gamma\gamma}}{\partial \gamma} = \frac{\partial B_3}{\partial \gamma}, \qquad (3.25)$$

$$B_{\beta\gamma} = 0, \; B_{\gamma\gamma} = B_3, \; B_1 = B_2 \; (\vartheta_1 = \vartheta_2, \vartheta_3 = 0). \qquad (3.26)$$

The symmetry relation (3.24) gives the following rules for the behaviour of the seven functions along the oblate ($\gamma = 60°$) axis:

$$\frac{\partial V}{\partial \delta} = 0 = \frac{\partial B_{\beta\beta}}{\partial \delta} = \frac{\partial B_{\gamma\gamma}}{\partial \delta} = \frac{\partial B_2}{\partial \delta}, \qquad (3.27)$$

$$B_{\beta\gamma} = 0, \; B_{\gamma\gamma} = B_2, \; B_1 = B_3 \; (\vartheta_1 = \vartheta_3, \vartheta_2 = 0), \qquad (3.28)$$

where δ is the magnitude of deformation perpendicular to the oblate axis.

The relations between $B_{\gamma\gamma}$ and the B_k (or the moment of inertia) are particularly important. Ignorance of these relations leads to many mistakes in many of the current versions of the collective model where one expands the collective Hamiltonian functions without obeying the symmetry rules (for example, Davydov and Chaban 1960), or where one multiplies the microscopically calculated inertial functions by more than one overall factor (for example, Dobaczewski et al. 1975).

Now let us consider some symmetries of the Hamiltonian as a whole which are useful for finding its solutions. First we need to consider the quantization of the generalized collective Hamiltonian of eq. (3.7) (or 3.6). Rewrite eq. (3.7) as

$$H_{coll} = V + T_{vib} + T_{rot}, \qquad (3.29)$$

$$T_{vib} = \tfrac{1}{2} B_{\beta\beta} \dot\beta^2 + B_{\beta\gamma} \dot\beta \dot\gamma + \tfrac{1}{2} B_{\gamma\gamma} \beta^2 \dot\gamma^2 , \qquad (3.30)$$

$$T_{rot} = \tfrac{1}{2}\left(\mathcal{I}_1 \omega_1^2 + \mathcal{I}_2 \omega_2^2 + \mathcal{I}_3 \omega_3^2 \right). \qquad (3.31)$$

The quantized collective Hamiltonian is written formally as

$$\hat H_C = V + \hat T_{vib} + \hat T_{rot} . \qquad (3.32)$$

The vibrational kinetic energy of (3.30) is quantized via the procedure employed in going from eq. (2.16) to (2.18). This gives (putting $\dot\alpha_1 = \dot\beta$, $\dot\alpha_2 = \dot\gamma$; see eq.(3.52) for F)

$$\hat T_{vib} = -\frac{\hbar^2}{2F}\left[\frac{\partial}{\partial\beta} F_{\beta\beta} \frac{\partial}{\partial\beta} + \frac{\partial}{\partial\beta} F_{\beta\gamma} \frac{\partial}{\beta\partial\gamma} \right.$$

$$\left. + \frac{\partial}{\beta\partial\gamma} F_{\beta\gamma} \frac{\partial}{\partial\beta} + \frac{\partial}{\beta\partial\gamma} F_{\gamma\gamma} \frac{\partial}{\beta\partial\gamma} \right], \qquad (3.33a)$$

$$\begin{pmatrix} F_{\beta\beta} & F_{\beta\gamma} \\ F_{\beta\gamma} & F_{\gamma\gamma} \end{pmatrix} = F\left(B_{\beta\beta} B_{\gamma\gamma} - B_{\beta\gamma}^2 \right)^{-1} \begin{pmatrix} B_{\gamma\gamma} & -B_{\beta\gamma} \\ -B_{\beta\gamma} & B_{\beta\beta} \end{pmatrix} . \qquad (3.33b)$$

The rotational kinetic energy (3.31) can be quantized in two different ways. (1) Write the angular frequencies in terms of the time-derivatives of the three Euler angles:

$$\omega_1 = \dot\phi \sin\psi - \dot\theta \sin\phi \cos\psi ,$$
$$\omega_2 = \dot\phi \cos\psi + \dot\theta \sin\phi \sin\psi , \qquad (3.34)$$
$$\omega_3 = \dot\theta \cos\phi + \dot\psi .$$

Then, use the same quantization procedure as that employed above. (2) Utilize the quantization conditions

$$\mathcal{I}_k \omega_k = \hbar I_k \qquad (k = 1,2,3), \qquad (3.35)$$

where I_k is the intrinsic-k-component of I. Both methods give the same result:

$$\hat T_{rot} = \frac{\hbar^2}{2} \sum_{k=1}^{3} \frac{I_k^2}{\mathcal{I}_k} . \qquad (3,36)$$

The quantized Hamiltonian (eq. 3.32 combined with eqs. 3.33 and 3.36) is rotationally invariant. In particular, it commutes with the rotation operators $R_k(\pi)$ which represent rotation of axes thru $180°$ while keeping the 'k' axis fixed:

$$\begin{aligned}R_1(\pi):\ & x_1 \to x_1,\ x_2 \to -x_2,\ x_3 \to -x_3, \\ R_2(\pi):\ & x_1 \to -x_1,\ x_2 \to x_2,\ x_3 \to -x_3, \\ R_3(\pi):\ & x_1 \to -x_1,\ x_2 \to -x_2,\ x_3 \to x_3.\end{aligned} \qquad (3.37)$$

Utilizing the anti-commutation relations

$$\left[H_{coll},\ R_k(\pi)\right] = 0 \qquad (k = 1,2,3), \qquad (3.38)$$

we require the wave functions to be eigenfunctions of R_k [in addition to being eigenfunctions of I^2, I_z (lab-z-axis), and Π (parity)]:

$$R_k(\pi) \mid IM > = r_k \mid IM > . \qquad (3.39)$$

Since R_k^2 gives +1 (-1) for even (odd) A, the quantum numbers r_k take the eigenvalus (Bohr & Mottelson 1975)

$$\begin{aligned} r_k &= \pm 1 \quad (A = \text{even}) \\ &= \pm i \quad (A = \text{odd}). \end{aligned} \qquad (3.40)$$

A wave function of angular momentum I must behave like a tensor of rank I. Hence the lab-system and the intrinsic-system components of the wavefunction must be related via

$$\mid IM > = \sum_K \mathcal{D}^I_{MK} \mid IK >, \qquad (3.41)$$

where K is the projection of $\vec{I}$ along the intrinsic-3-axis. Using the properties of $\mathcal{D}$-functions and the general condition $r_1 r_2 r_3$ = fixed for a given system, eqs. (3.39, 3.41) are combined to yield (for nuclei with A = even)

$$\mid IM > = \sum_{\substack{K=0,2,4,\ldots(r_3=+1) \\ K=1,3,5,\ldots(r_3=-1)}} A_{IK}\, \Phi^I_{MK}, \qquad (3.42a)$$

where
$$\Phi^I_{MK} = \sqrt{\frac{2I+1}{16\pi^2(1+\delta_{K0})}} \left[\mathcal{D}^I_{MK} + r_2(-1)^{I+K} \mathcal{D}^I_{M,-K}\right], \qquad (3.42b)$$

and A_{IK} is proportional to $|IK>$. Note that a normalization factor has been introduced in eq. (3.42b) so as to make Φ obey the normalization condition

$$\iiint \left|\Phi_{MK}^{I}\right|^2 \sin\theta \, d\theta \, d\phi \, d\psi = 1. \qquad (3.43)$$

The $\mathcal{D}$-functions are normalized via the equation

$$\iiint \left|\mathcal{D}_{MK}^{I}\right|^2 \sin\theta \, d\theta \, d\phi \, d\psi = \frac{8\pi^2}{2I+1}. \qquad (3.44)$$

For the low-energy states of even-even nuclei, the r_k values are chosen to be

$$r_1 = r_2 = r_3 = +1 . \qquad (3.45)$$

This choice is appropriate for the seniority zero states where all nucleons are paired (not necessarily to J = 0) in such a way that each single-particle state is occupied with the same probability as the corresponding tim-reserved state. This time-reversal symmetry has been identified as the crucial reason for the occurrence of "rotational" type features (a sequence of levels with I = 0,2,4,... connected by large B(E2) values and with little mixing with the other levels) in practically all even-even nuclei (Kumar 1975 c).

The collective Schrödinger Equation is written in the intrinsic system as

$$\hat{H}_c |IM> = (V + \hat{T}_{vib} + \hat{T}_{rot}) |IM> = E_I |IM> , \qquad (3.46)$$

where $\hat{T}_{vib}$, $\hat{T}_{rot}$, $|IM>$ are given by eqs. 3.33, 3.36, 3.42, respectively. This represents a set of coupled partial differential equations since K is not a good quantum number in general. In fact, the degree of non-linearity is given by the number of K values allowed for each I. With the symmetry implied by eq. (3.45), the allowed values are given in Table 3 for $I \leq 6$. Note that there is no I = 1 state in this table since the corresponding wave function vanishes. Also, the wave function vanishes for K = 0 if I = odd. The general rules for the allowed K-values are:

$$K = 0,2,4,\ldots,I \text{ if } I = \text{even},$$
$$= 2,4,\ldots,(I-1) \text{ if } I = \text{odd} . \qquad (3.47)$$

These rules lead to the band sequences,

$$K = 0: \quad I = 0,2,4,6,\ldots$$
$$K = 2; \quad I = 2,3,4,5,6,\ldots \qquad (3.48)$$
$$K = 4; \quad I = 4,5,6,\ldots$$

Note that these sequences follow from the symmetry ($r_1 = r_2 = r_3 = +$) rather than any assumption about the vibrational or rotational nature of the nucleus.

As regards the β-γ-dependence of the allowed wave functions, the main reqirements are determined via the same procedure as that employed above for the Hamiltonian functions. A wave function for angular momemtum I can be expressed as a tensor of rank I obtained via the coupling of one or more quadrupole tensors:

$$|IM\rangle = [\alpha_2 \otimes \alpha_2 \cdots \otimes \alpha_2]_{IM} \,. \tag{3.49}$$

The β-γ-K-dependence of the allowed basic tensors for different I is determined by transforming to the intrinsic system. Such basic tensors for $I \leq 6$ are listed in Table 3.

These basic tensors are employed to solve the vibrational Hamiltonian (see sec. III.C), as well as the general Hamiltonian (next section).

III.B. Exact Treatment of Rotation-Vibration Coupling

(Numerical Solution Method)

A numerical method of solving the collective Schrödinger Equation (CSE) was proposed and applied to the transitional osmium nuclei (Kumar 1963). This method has been improved over the years (Kumar and Baranger 1967, Kumar 1974, Kumar 1979a). The latest version is discussed below.

Recall that the CSE is given by Eq. (3.46) where $\hat{T}_{vib}$, $\hat{T}_{rot}$, $|IM\rangle$ are given by Eqs. 3.33, 3.36, 3.42, respectively. The Euler-angle-dependent part of the wave function, ϕ_{MK}^I, is known. Our objective is to determine the β-γ-K-dependent wave function, $A_{IK}(\beta,\gamma)$.

Let us first consider the normalization of the wave function. We have here a problem where the mass (the mass-matrix $B_{\mu\nu}$) depends on the coordinates. The quantization rule of Pauli, which was employed to quantize the classical collective Hamiltonian, tells us that the volume element is given by

$$d\tau = D^{\frac{1}{2}} \prod_\mu d\alpha_\mu \,, \tag{3.50}$$

where D is the determinant of the mass-matrix (see Eq. 2.18). In the intrinsic system (β,γ,φ,θ,ψ), the determinant is given by

$$D = F^2 \sin^2\theta, \tag{3.51}$$

where $F = \left|\left(B_{\beta\beta}B_{\gamma\gamma} - B_{\beta\gamma}^2\right) \mathcal{I}_1 \mathcal{I}_2 \mathcal{I}_3\right|^{\frac{1}{2}} \beta \,.$ (3.52)

The five-dimensional integrals can be written as products of a three-dimensional integral over ϕ,θ,ψ (see Eq. 3.43 for the normalization of Φ^I_{MK}) and a two dimensional integral over β,γ. The later take the form

$$\int_0^\infty d\beta \int_0^{\pi/3} d\gamma A_{I'K'}(\beta,\gamma) \, \Theta(\beta,\gamma) \, A_{IK}(\beta,\gamma) \, F, \qquad (3.53)$$

where Θ is a β-γ-dependent operator.

The wave function $|IM\rangle$ of Eq. (3.42) is normalized in such a way that the ϕ-θ-ψ-dependent part and the β-γ-dependent part are normalized separately in their own spaces. While the former is normalized via Eq. (3.43), the later is normalized via the following condition:

$$\sum_K \int_0^\infty d\beta \int_0^{\pi/3} d\gamma \, F \, A_{IK}^2 = 1 \,. \qquad (3.54)$$

Using the general relation (3.13), the eq. (3.52) can be rewritten as

$$F = 2|\beta^4 \sin 3\gamma| \left| \left(B_{\beta\beta} B_{\gamma\gamma} - B_{\beta\gamma}^2\right) B_1 B_2 B_3 \right|^{\frac{1}{2}}. \qquad (3.55)$$

The "weight-factor" F vanishes for the spherical shape ($\beta = 0$), for the prolate shapes ($\gamma = 0°$), and for the oblate shapes ($\gamma = 60°$). Moreover, the mass parameters ($B_{\beta\beta}$, $B_{\beta\gamma}$, $B_{\gamma\gamma}$, B_1, B_2, B_3) vary with β and γ by several orders of magnitude (Kumar 1975 a). Hence, the standard procedure of expanding the desired solution in a basis of known, orthonormal states (for instance, those of the vibrational model) is not suitable for the most general treatment of the present problem. Instead, the following steps are employed.

In the first step, the three rotational degrees of freedom (ϕ,θ,ψ) are eliminated analytically by employing the known properties of the angular momentum operators and the $\mathcal{D}$-functions. Thus, the five-dimensional CSE of eq. (3.46) is reduced to a two-dimensional non-linear integro-differential equation:

$$E_I \langle IM | IM \rangle = \langle IM | \hat{H}_c | IM \rangle$$

$$= \int_0^\infty \beta d\beta \int_0^{\pi/3} d\gamma \, F \sum_K \left[V A_{IK}^2 + A_{IK} \hat{T}_{vib} A_{IK} \right.$$

$$\left. + \sum_{K'} A_{IK} T_{KK'} A_{IK'} \right], \qquad (3.56)$$

where $\hat{T}_{vib}$ is the differential operator of eq. (3.33) and $T_{KK'}$ is the matrix of the rotational operator of eq. (3.36) with respect to the rotational wave function of eq. (3.42b).

The matrix $T_{KK'}$ is determined by employing the following rules for the angular momentum operators:

$$I_3 \mathcal{D}^I_{MK} = K \mathcal{D}^I_{MK}, \qquad (3.57a)$$

$$I_\pm \mathcal{D}^I_{MK} = \sqrt{(I \pm K)(I \mp K + 1)}\, \mathcal{D}^I_{M,K \mp 1}, \qquad (3.57b)$$

where $I_\pm = I_1 \pm i I_2$. $\qquad (3.57c)$

Note that according to eq. (3.57b), I_+ (I_-) lowers (raises) the K value by one unit. This is the opposite of the usual rule. This occurs because the commutation rule for the intrinsic components of $\vec{I}$ has an extra minus sign (Bohr & Mottelsen 1969),

$$\left[I_i, I_j \right] = -i I_k \quad (i,j,k \text{ are cyclic}). \qquad (3.58)$$

Non-zero matrix elements of the matrix $T_{KK'}$ are given by:

$$T_{KK} = aI(I+1) + bK^2, \qquad (3.59a)$$

$$T_{K,K+2} = T_{K+2,K} = c\left[(1+\delta_{K0})(I-K-1)(I-K)(I+K+1)(I+K+2)\right]^{\frac{1}{2}}, \quad (3.59b)$$

where $a = \dfrac{\hbar^2}{4}\left(\dfrac{1}{\mathcal{J}_1} + \dfrac{1}{\mathcal{J}_2}\right)$, $\qquad (3.59c)$

$b = \dfrac{\hbar^2}{2\mathcal{J}_3} - a$, $\qquad (3.59d)$

and $c = \dfrac{\hbar^2}{8}\left(\dfrac{1}{\mathcal{J}_1} - \dfrac{1}{\mathcal{J}_2}\right)$. $\qquad (3.59e)$

In the second step of the present method, integrations over β and γ are performed by employing a numerical approximation and a triangular $\beta - \gamma$ mesh (see Fig. 8). According to this approximation (Kumar & Baranger 1967), the double integral of any function $f(\beta,\gamma)$ is given by

$$\int \beta d\beta \int d\gamma\, f(\beta,\gamma) = \frac{s^2}{\sqrt{3}} \sum_i w_i f_i, \qquad (3.60)$$

where s is the mesh size (s = 0.05 for the mesh of Fig. 8, with β_{max} = 0.8), f_i is the numerical value of the function at the mesh point 'i', and w_i is a weight-factor.

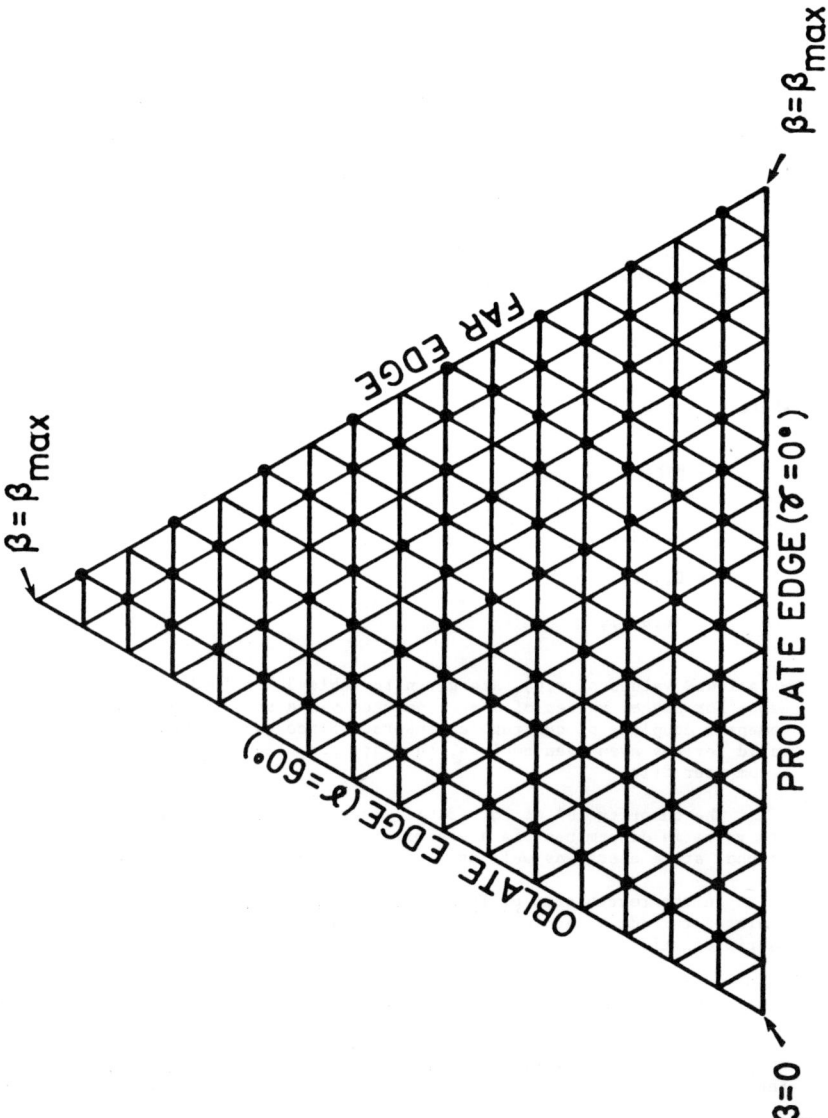

Fig. 8. The Beta-Gamma Mesh Used for Integrations Over β and γ.

Because of the symmetries of the equilateral triangle, the weight-factor vanishes for about 20% of the mesh points. Furthermore, the function f_i (which always includes the factor of eq. 3.55 in the five-dimensional problem under consideration) vanishes along the prolate as well as the oblate edge of the triangular mesh. This fact allows us to reduce the number of mesh points, at which the function f must be known, by another 20%. In most of the calculations based on this method, each edge of the triangle is divided into 16 steps. The corresponding β-γ-mesh contains 153 points. However, the various microscopic functions $(V, \mathcal{J}_1, \mathcal{J}_2, \ldots)$, which determine f, need to be computed at only 92 of these mesh points. Values for the spherical shape or the axially symmetric (prolate or oblate) shapes never enter the solution of the CSE in the present method!

In the third step of the method, the wave functions A_{IK} are expanded as follows:

$$A_{IK} = \sum_p a_p P_{pIK}, \qquad (3.61a)$$

where $P_{pIK} = D^{2r}(D^3\cos 3\gamma)^s D^v g_{IK}^v(\gamma)\exp\left(-\frac{1}{2}D^2\right);$ (3.61b)

$$D = \beta/b; \qquad (3.61c)$$

$$r = 0,1,2,\ldots; \qquad (3.61d)$$

$$s = 0,1,2,\ldots; \qquad (3.61e)$$

and $\quad p = (r,s,v).$ (3.61f)

The allowed values of v, g_{IK}^v are given in Table 3 for $I = 0-6$. Components for higher values of I are computed via a program for the angular momentum coupling of quadrupole tensors. Since these components are valid for any even-even nucleus, they are computed once and stored on a computer file.

The quantity 'b' of eq. (3.61c) represents the gaussian range. Its value is determined via a minimization of the variation of the ground state energy as well as the first excited 0^+ state energy.

In the fourth step of the method, equations (3.60, 3.61) are employed to reduce the integro-differential equation (3.56) to the matrix equation.

$$\sum_q H_{pq} a_{q\alpha} = E_{\alpha I} \sum_q N_{pq} a_{q\alpha}, \qquad (3.62)$$

where α represents different solutions (with the same I) and N_{pq} is the norm-matrix. This norm-matrix is not an unity-matrix in the

Table 3. Allowed K values and basic tensor components of collective wave functions. The quantity v equals the number of angular-momentum-coupled-quadrupole tensors.

I	Matrix Size		v	K	g_{IK}^v
	K.B.[a]	K.[b]			
0	120	16	0	0	1
2	272	30	1	0	$\cos \gamma$
				2	$\sin \gamma$
			2	0	$\cos 2\gamma$
				2	$-\sin 2\gamma$
3	120	16	3	2	$\sin 3\gamma$
4	408	42	2	0	$7 + 5 \cos 2\gamma$
				2	$\sqrt{60} \sin 2\gamma$
				4	$\sqrt{35} (1 - \cos 2\gamma)$
			3	0	$5 \cos \gamma + 7 \cos 3\gamma$
				2	$- \sqrt{60} \sin \gamma$
				4	$\sqrt{35} (- \cos \gamma + \cos 3\gamma)$
			4	0	$7 + 5 \cos 4\gamma$
				2	$- \sqrt{60} \sin 4\gamma$
				4	$\sqrt{35} (1 - \cos 4\gamma)$
5	240	30	4	2	$\cos \gamma \sin 3\gamma$
				4	$\sin \gamma \sin 3\gamma$
			5	2	$\cos 2\gamma \sin 3\gamma$
				4	$- \sin 2\gamma \sin 3\gamma$

Table 3 (continued)

I	Matrix Size K.B.[a]	K.[b]	v	K	g_{IK}^{v}
6	544	52	3	0	$\sqrt{10}\,(21\cos\gamma + 3\cos 3\gamma)$
				2	$\sqrt{7}\,(15\sin\gamma + 11\sin 3\gamma)$
				4	$\sqrt{70}\,(6\sin\gamma\sin 2\gamma)$
				6	$\sqrt{385}\,(3\sin\gamma - \sin 3\gamma)$
			4	0	$\sqrt{10}\,(3 + 14\cos 2\gamma + 7\cos 4\gamma)$
				2	$\sqrt{7}\,(-10\sin 2\gamma + 5\sin 4\gamma)$
				4	$\sqrt{70}\,(-3 + 2\cos 2\gamma + \cos 4\gamma)$
				6	$\sqrt{385}\,(-2\sin 2\gamma + \sin 4\gamma)$
			5	0	$\sqrt{10}\,(14\cos\gamma + 3\cos 3\gamma + 7\cos 5\gamma)$
				2	$\sqrt{7}\,(10\sin\gamma - 11\sin 3\gamma - 5\sin 5\gamma)$
				4	$\sqrt{70}\,(2\cos\gamma - 3\cos 3\gamma + \cos 5\gamma)$
				6	$\sqrt{385}\,(2\sin\gamma + \sin 3\gamma - \sin 5\gamma)$
			6	0	$\sqrt{10}\,(21\cos 2\gamma + 3\cos 6\gamma)$
				2	$-\sqrt{7}\,(15\sin 2\gamma + 11\sin 6\gamma)$
				4	$\sqrt{70}\,(6\sin 2\gamma\sin 4\gamma)$
				6	$\sqrt{385}\,(-3\sin 2\gamma + \sin 6\gamma)$

[a] Kumar and Baranger 1967.

[b] Kumar 1979a.

present problem since the basis functions, P_{pIK} of eqs. (3.61), are not orthonormal. Hence, the norm-matrix is first diagonalized by solving the standard eigenvalue equation,

$$\sum_q N_{pq} b_{qi} = n_i b_{pi} . \qquad (3.63)$$

Then, the modified hamiltonian-matrix is calculated and solved via the equations

$$\sum_j h_{ij} C_{j\alpha} = E_{\alpha I} C_{i\alpha} , \qquad (3.64a)$$

$$h_{ij} = \frac{1}{\sqrt{n_i n_j}} \sum_{pq} b_{pi} H_{pq} b_{qj} , \qquad (3.64b)$$

and $\quad a_{p\alpha} = \sum_i \frac{b_{pi} C_{i\alpha}}{\sqrt{n_i}} . \qquad (3.64c)$

Eigenvalues of the matrix equation (3.64a), $E_{\alpha I}$, give the collective energies for each I value. The eigenvalues, $C_{i\alpha}$, of the same equation are combined with eqs. (3.64c) and (3.61) to obtain the collective wave functions A_{IK} (abbreviation for $A_{\alpha IK}$).

These wave functions are employed to compute the B(E2) and B(M1) values, quadrupole and magnetic moments, E2/M1 and E0/E2 mixing ratios, isomer shifts,... The general relations given previously (Kumar 1975a) are employed.

C. Spherical (Vibrational) Limit

In the lab system, the collective Hamiltonian in the spherical limit is given by eq. (1.1). This represents harmonic, uncoupled vibrations in the five-dimensional space (α_μ with $\mu = 0, \pm 1, \pm 2$). The most straightforward way of solving this problem is to replace the "coordinates" α_μ and the "momenta" $\Pi_\mu = B\dot\alpha_\mu$ by boson operators C_μ^+, C_μ (following Messiah 1962)

$$\alpha_\mu = \left(\frac{\hbar}{2B\omega}\right)^{\frac{1}{2}} \left(C_\mu^+ + C_\mu\right) , \qquad (3.65a)$$

$$\Pi_\mu = i\left(\frac{\hbar B\omega}{2}\right)^{\frac{1}{2}} \left(C_\mu^+ - C_\mu\right) . \qquad (3.65b)$$

Then, eq. (1.1) becomes

$$\hat{H}_c = \hbar\omega \sum_\mu \left(c_\mu^+ c_\mu + \frac{1}{2} \right), \tag{3.66a}$$

where $\omega = \sqrt{C/B}$. (3.66b)

The corresponding eigenfunctions are given by

$$|NIM> = [c^+ \otimes c^+ \ldots \otimes c^+]_{IM} \, |->, \tag{3.67}$$

where N is the number of bosons (phonons) coupled to I,M (each boson carries 2 unit of angular momentum) and $|->$ represents the vacuum (N=0) state. The corresponding eigenvalues are (which can be checked by employing the general commutation rules for boson operators)

$$E_N = (N + 5/2)\hbar\omega \quad (N = 0,1,2,\ldots). \tag{3.68}$$

The angular momentum values allowed for a given number of phonons are determined by the angular momentum coupling rules. The vibrational spectrum for N=0,1,2,3 is shown in Fig. 9.

In 1952 Bohr showed that the same problem can be solved <u>exactly</u> in the intrinsic system $(\beta,\gamma,\phi,\theta,\psi)$. The solution is more complicated, but there are no theoretical difficulties such as those imagined by von Bernus et al. (1974).

The general transformation of the collective Hamiltonian from the lab system to the intrinsic system has been discussed above. Now we consider its special case in the vibrational limit. With the potential energy given by $V = \frac{1}{2}C\beta^2$, and the inertial functions by eqs. (3.12), the quantized Hamiltonian of eqs. (3.32, 3.33, 3.36) becomes

$$H_{coll} = \frac{1}{2}C\beta^2 - \frac{\hbar^2}{2B}\left(\frac{1}{\beta^4}\frac{\partial}{\partial\beta}\beta^4\frac{\partial}{\partial\beta} + \frac{1}{\beta^2 \sin 3\gamma}\frac{\partial}{\partial\gamma}\sin 3\gamma\frac{\partial}{\partial\gamma}\right)$$

$$+ \frac{\hbar^2}{2B\beta^2}\sum_{k=1}^{3}\frac{I_k^2}{4\sin^2(\gamma-\frac{2}{3}\pi k)}. \tag{3.69}$$

The general eigenfunctions of eq. (3.42) are written as

$$|NIM> = \sum_{K=0,2,4,\ldots} A_{NIK}(\beta,\gamma) \, \Phi_{MK}^{I}(\phi\theta\psi), \tag{3.70}$$

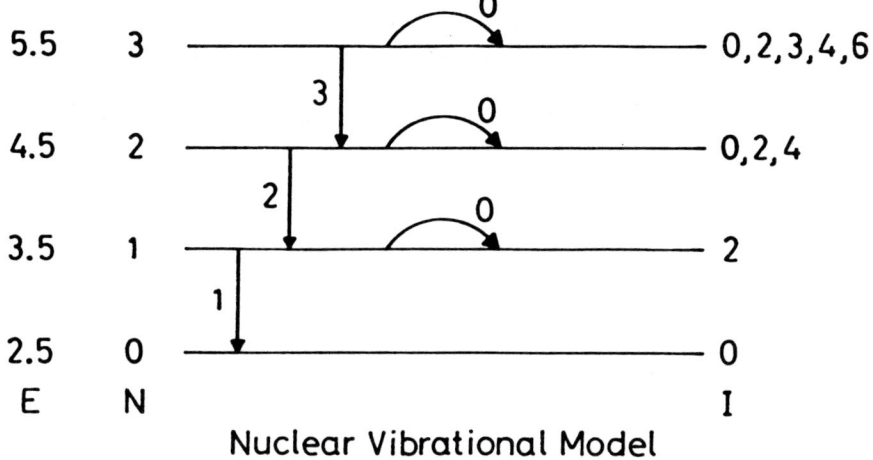

Fig. 9. Ideal vibrational spectrum with the associated B(E2) values (↓) and spectroscopic quadrupole moments (~). The B(E2) values represent a sum over different I-values allowed for the final N-value, and are in units of B(E2; 2→0). The energy E is in units of $(E_2 - E_0)$.

where N is the number of quadrupole tensors employed to construct the state $|NIM\rangle$. In principle, one could start with a wave function like that of eq. (3.67) where $C_{2\mu}^+$ is replaced by α_μ, and then make a transformation to the intrinsic system via the $\mathcal{D}$-functions. However, it is not very pleasant to have to couple a long string of $\mathcal{D}$-functions. Also, it is more instructive to solve this problem in a manner similar to that employed for the harmonic oscillator in the r,θ,ϕ space. The spherical harmonics is replaced by the $\mathcal{D}$-function, while the radial wave function is replaced by $A_{NIK}(\beta,\gamma)$. The boundary conditions are included via a Gaussian which vanishes as $\beta \to \infty$ and via the basic tensors of Table 3. Thus, the trial wave function is given by eqs.(3.61).

The mixing coefficients of eq. (3.61a) are determined analytically for the present case of harmonic quadrupole vibrations. Since the mass parameters are independent of deformation,

$$B_{\beta\beta} = B_{\gamma\gamma} = B_1 = B_2 = B_3 = B, \quad B_{\beta\gamma} = 0; \tag{3.71}$$

the weight-factor (see eq. 3.55) becomes

$$F = 2B^{5/2} \beta^4 \sin 3\gamma. \tag{3.72}$$

In analogy with the oscillator problem, we make the coordinate transformation

$$D = \beta/b, \tag{3.73a}$$

where the range parameter 'b' is given by

$$b = \sqrt{\frac{\hbar}{B\omega}}, \quad \omega = \sqrt{\frac{C}{B}}. \tag{3.73b}$$

Then, the integrals of eqs. (3.53) and (3.54) become

$$\int d\tau \; A_{N'I'K'}(D,\gamma) \ominus (D,\gamma) A_{NIK}(D,\gamma), \tag{3.74a}$$

and

$$\int d\tau \sum_K A_{NIK}^2 = 1, \tag{3.74b}$$

where $d\tau = 2(\hbar/\omega)^{5/2} D^4 \sin 3\gamma \, dD d\gamma.$ \hfill (3.74c)

The various integrals can be expressed as products of a D-integral of type (n=integer)

$$I_n = \frac{8}{3\sqrt{\pi}} \int_0^\infty D^{2n+4} \exp(-D^2) dD = \frac{\left(n+\frac{3}{2}\right)!}{\frac{3}{2}!}, \tag{3.75a}$$

and a γ-integral of type

$$C_n = \frac{3}{2} \int_0^{\pi/3} \cos n\gamma \sin 3\gamma \, d\gamma = \frac{9\left(1+\cos \frac{n\pi}{3}\right)}{2(9-n^2)} \quad (n \neq 3), \qquad \text{3.75b}$$

or of type

$$S_n = \frac{3}{2} \int_0^{\pi/3} \sin n\gamma \sin 3\gamma \, d\gamma = \frac{9 \sin \frac{n\pi}{3}}{2(9-n^2)} \quad (n \neq 3). \qquad (3.75c)$$

For the case of $n = 3$, we have

$$C_3 = 0, \quad S_3 = \pi/4. \qquad (3.75d)$$

The energy integral of eq. (3.56) is written as

$$E_{NI} = V_{NI} + K_{NI} + R_{NI}, \qquad (3.76a)$$

where $V_{NI} = \int d\tau \sum_K A_{NIK}^2 \left(\frac{1}{2} C \beta^2\right) = \frac{1}{2} \hbar\omega \int d\tau \sum_K A_{NIK}^2 D^2, \qquad (3.76b)$

$$K_{NI} = -\frac{1}{2} \hbar\omega \int d\tau \sum_K A_{NIK} \left(\frac{1}{D^4} \frac{\partial}{\partial D} D^4 \frac{\partial A_{NIK}}{\partial D}\right.$$

$$\left. + \frac{1}{D^2 \sin 3\gamma} \frac{\partial}{\partial \gamma} \sin 3\gamma \frac{\partial A_{NIK}}{\partial \gamma}\right), \qquad (3.76c)$$

$$R_{NI} = \frac{1}{2} \hbar\omega \int \frac{d\tau}{D^2} \sum_{KK'} A_{NIK} t_{KK'} A_{NIK'}, \qquad (3.76d)$$

$$t_{KK'} = \frac{2B\beta^2}{\hbar^2} T_{KK'} = \frac{2D^2}{\hbar\omega} T_{KK'}, \qquad (3.76e)$$

and the wave functions A_{NIK} are assumed to be normalized.

Because of the extra complication due to the extra degrees of freedom compared to the three-dimensional oscillator, it is not possible to give a simple analytical expression for all the solutions of the 5-D oscillator. But we can make a reasonable guess for the solution and then show that the solution gives the correct energy (eq. 3.68) and that it is orthogonal to the lower energy solutions.

The reasonable guess is based on eqs. (3.61) and on the observation that the N-phonon solution can be written as the product of N bosons (see eqs. 3.65, 3.67). Therefore, the maximum power of D (or β) occuring in the solution equals N. Also, the Hamiltonian is even in β (see eq. 3.69). Therefore, the solution is either even or odd in β depending on whether N is even or odd. Hence, the allowed values of r,s of eqs. (3.61) are restricted (for the harmonic case considered here) by the relation

$$2r + 3s + v = N, N-2, N-4,\ldots \quad (0 \text{ or } 1). \tag{3.77}$$

The allowed values of v are restricted by (see Table 3 for $I \leq 6$)

$$v = \tfrac{1}{2}I, \tfrac{1}{2}I+1,\ldots, I \quad (I = \text{even}), \tag{3.78a}$$

$$= \tfrac{1}{2}(I+3), \tfrac{1}{2}(I+5),\ldots, I \quad (I = \text{odd}). \tag{3.78b}$$

Now, we consider the lowest few solutions of the 5-D vibrational model. The ground state is given by ($N = r = s = v = I = 0$)

$$A_{000} = \nu_0 \exp(-\tfrac{1}{2}D^2), \tag{3.79a}$$

$$\nu_0 = \left(\frac{4\omega^5}{\pi\hbar^5}\right)^{1/4}. \tag{3.79b}$$

The corresponding potential energy $\langle V \rangle$ is given by (using eqs. 3.76b, 3.75a)

$$V_{00} = \tfrac{1}{2}\hbar\omega\, I_1 = \tfrac{5}{4}\hbar\omega, \tag{3.79c}$$

and the corresponding vibrational energy $\langle T_{vib} \rangle$ is given by

$$K_{00} = -\tfrac{1}{2}\hbar\omega \int d\tau\, (D^2 - 5)\, A_{000}^2$$

$$= -\tfrac{1}{2}\hbar\omega\, (I_1 - 5) = \tfrac{5}{4}\hbar\omega. \tag{3.79d}$$

There is no rotational energy for $I = 0$, of course. Note that the vibrational energy contributes one-half ot the total ground state energy (of $\tfrac{5}{2}\hbar\omega$), as is typical for an oscillator potential. The ground state energy equals the energy of zero-point-motion which is $\tfrac{5}{2}\hbar\omega$ for the five-dimensional motion.

The one-phonon wave function is given by ($N = v = 1$, $I = 2$, $r = s = 0$)

$$\begin{pmatrix} A_{120} \\ A_{122} \end{pmatrix} = \sqrt{\tfrac{2}{5}}\, DA_{000} \begin{pmatrix} \cos\gamma \\ \sin\gamma \end{pmatrix}. \tag{3.80a}$$

The total wave function includes the function Φ_{MK}^2 (see eqs. 3.42 and 3.45) and is orthogonal to the ground state because of the Φ-function. The corresponding potential energy is given by

$$V_{12} = \tfrac{1}{2}\hbar\omega\left(\tfrac{2}{5} I_2\right) = \tfrac{7}{4}\hbar\omega, \tag{3.80b}$$

while the vibrational energy is given by

$$K_{12} = -\frac{1}{2}\hbar\omega \left(\frac{2}{5}\right) \int d\tau \, A_{000}^2 \, (3 - 7D^2 + D^4)$$

$$= -\frac{1}{5}\hbar\omega \, (3 - 7I_1 + I_2) = \frac{23}{20}\hbar\omega . \qquad (3.80c)$$

The rotational-matrix for $I = 2$ is given by (using eqs. 3.12, 3.59 and 3.76)

$$t_{00} = 6A, \quad t_{02} = t_{20} = 4C\sqrt{3}, \quad t_{22} = 6A + 4B, \qquad (3.80d)$$

where
$$A = \frac{\sin^2\gamma + 3\cos^2\gamma}{(\sin^2\gamma - 3\cos^2\gamma)^2} , \qquad (3.80e)$$

$$B = \frac{1}{4\sin^2\gamma} - A , \qquad (3.80f)$$

and
$$C = -\frac{\sqrt{3}\sin\gamma \cos\gamma}{(\sin^2\gamma - 3\cos^2\gamma)^2} . \qquad (3.80g)$$

The relations for A,B,C are valid for any I-value of the oscillator model.

The rotational energy for the one-phonon state is given by

$$R_{12} = \frac{1}{2}\hbar\omega \left(\frac{2}{5}\right) \int d\tau \, A_{000}^2 \, (t_{00} \cos^2\gamma + 2t_{02} \sin\gamma \cos\gamma + t_{22} \sin^2\gamma)$$

$$= \frac{1}{5}\hbar\omega \int d\tau \, A_{000}^2 \, (3) = \frac{3}{5}\hbar\omega . \qquad (3.80h)$$

The sum of the three contributions of eqs. (3.80b, c, h) gives the total energy of the one-phonon state to be $\frac{7}{2}\hbar\omega$. Note that while the potential energy contributes 50%, the vibration contributes 33%, and the rotation contributes 17% of the total energy. If we consider differences with respect to the ground state, we find that the respective potential, vibrational, rotational contributions to the excitation energy $\hbar\omega$ of the first 2+ state are: 50%, -10%, and 60%. Thus, it is wrong to attribute all the excitation energy of the first 2+ state to rotations of all even-even nuclei. Only 60% of it comes from rotations in the limit of harmonic vibrations around the spherical shape. The corresponding "effective" moment of inertia should be 5 times $\hbar^2/E_2$ instead of 3 times.

While no extra effort was involved in making the first excited state orthogonal to the lower states (their angular momenta being different), the orthogonalization is essential for the second and higher states with the same symmetry quantum numbers. In order to illustrate this point, we consider the two-phonon wave function with $I = 0$ ($N = 2$, $v = s = 0$, $r = 0,1$). The trial wave function is given by

$$A_{00} = (a_0 + a_1 D^2) \, A_{000}. \qquad (3.81a)$$

The coefficients a_0, a_1 are determined by the normalization condition, which gives

$$1 = a_0^2 + 2a_1 a_0 I_1 + a_1^2 I_2 = a_0^2 + 5a_1 a_0 + \frac{35}{4} a_1^2 , \qquad (3.81b)$$

and by the condition of orthogonality to the ground state, which gives

$$0 = a_0 + a_1 I_1 = a_0 + \frac{5}{2} a_1 . \qquad (3.81c)$$

On combining eqs. (3.81a - c), we find the solution to be

$$A_{200} = \sqrt{\frac{5}{2}} \left(1 - \frac{2}{5} D^2\right) A_{000} . \qquad (3.81d)$$

The corresponding potential energy is given by

$$V_{20} = \frac{5}{2} \left(I_1 - \frac{4}{5} I_2 + \frac{4}{25} I_3\right) \frac{1}{2}\hbar\omega = \frac{9}{4}\hbar\omega , \qquad (3.81e)$$

and the vibrational energy by

$$K_{20} = -\frac{1}{2}\hbar\omega \left(\frac{5}{2}\right)\left(-9 + \frac{41}{5} I_1 - \frac{56}{25} I_2 + \frac{4}{25} I_3\right) = \frac{9}{4}\hbar\omega . \qquad (3.81f)$$

The sum of the two gives the total energy, $\frac{9}{2}\hbar\omega$, of the 0'+ state of the vibrational model (the rotational energy being zero for $I = 0$). Again, half of the toal comes from $\langle V \rangle$ and the other half from $\langle \hat{T}_{vib} \rangle$.

Wave functions for the higher states can be obtained in proceeding in the above manner. The basic functions given in Table 3 provide a good starting point.

It is also possible to give analytical, recursion type relations for the solutions of the vibrational model (see, for example, Turner and Kishimoto 1973). However, such relations mask the simple properties of the lowest energy solutions. Hence, the lowest few solutions are given in Table 4. Note that the quantum number v becomes a good quantum number in the present case. Hence, the γ-dependence (except for some powers of cos3γ for s $\neq$ 0) is essentially given by the functions $g_{IK}^v(\gamma)$, given in Table 3. The table 4 also gives the expectation values of V, $\hat{T}_{vib}$ and $\hat{T}_{rot}$. Note that this does not imply the separation of rotations and vibrations. The two are, in general, strongly coupled thru the summation over K-values (see Eq. 3.70). However, the Hamiltonian can be written as the sum of three distinct terms whose expectation values are given in Table 4. The table also gives the percentage contribution of each K-component to the normalization integral of Eq. (3.74).

The root-mean-square value of deformation is given for the vibrational model by

$$\beta_{rms}^2(N) = \langle \beta^2 \rangle = \frac{2}{C} \langle V \rangle = \left(N + \frac{5}{2}\right)\frac{\hbar\omega}{C} = \left(N + \frac{5}{2}\right) b^2 . \qquad (3.82)$$

Although the potential minimum corresponds to the spherical shape, the rms

Table 4. Wave Functions and Some Expectations Values for the $N \leq 3$ States of the Vibrational Model.

N	I	v	$\langle V \rangle$ [a]	$\langle \hat{T}_{vib} \rangle$ [a]	$\langle \hat{T}_{rot} \rangle$ [a]	$A_{NIK}/(A_{000} g^v_{IK})$ [b]	K	c^c_K
0	0	0	1.25	1.25	0.00	1	0	100.0
1	2	1	0.50	-0.10	0.60	$\left(\frac{2}{5}\right)^{\frac{1}{2}} D$	0	72.5
							2	27.5
2	0	0	1.00	1.00	0.00	$-\left(\frac{5}{2}\right)^{\frac{1}{2}}(1 - \frac{2}{5} D^2)$	0	100.0
	2	2	1.00	0.14	0.86	$-\frac{2}{(35)^{\frac{1}{2}}} D^2$	0	33.9
							2	66.1
	4	2	1.00	-0.19	1.19	$\frac{1}{6(35)^{\frac{1}{2}}} D^2$	0	61.8
							2	27.5
							4	10.7
3	0	0	1.50	1.50	0.00	$-\left(\frac{8}{105}\right)^{\frac{1}{2}} D^3 \cos 3\gamma$	0	100.0
	2	1	1.50	0.90	0.60	$-\left(\frac{7}{5}\right)^{\frac{1}{2}} (D - \frac{2}{7} D^3)$	0	72.5
							2	27.5
	3	3	1.50	0.00	1.50	$\frac{2}{(105)^{\frac{1}{2}}} D^3$	2	100.0
	4	3	1.50	0.21	1.29	$-\frac{1}{3(385)^{\frac{1}{2}}} D^3$	0	44.3
							2	18.8
							4	37.0
	6	3	1.50	-0.26	1.76	$\frac{1}{4(105 \times 143)^{\frac{1}{2}}} D^3$	0	54.2
							2	31.0
							4	9.6
							6	5.2

[a] The table gives the absolute values for the ground state, and the relative values (e.g. $\langle V \rangle - \langle V \rangle gs$) for the excited states, in units of $\hbar\omega$.

[b] The quantity A_{000} is given by Eqs. (3.79 a,b) of the text. The γ-dependent functions g^v_{IK} are listed in Table 3.

[c] Percentage contribution to the normalization integral.

value of deformation is non-zero. Furthermore, this value increases with N, the number of vibrational quanta. However, the model nucleus does not become necessarily more "rotational" with increasing N. Consider the yrast states in this model, the states with the lowest energy for $I = 0, 2, 4, \ldots$. These states have $I = N/2$ and the % $K = 0$ component is given by (see Table 4) 100, 73, 62, 54, ... for $I = 0, 2, 4, 6, \ldots$. Thus, although the lowest few yrast states are largely $K = 0$ states, the K-purity decreases even as β_{rms} increases with N.

The low-energy spectrum of the vibrational model is shown in Fig. 9. The absolute energy of each state, E, is given in units of $\hbar\omega$, excitation energy of the first 2+ state . As is typical of a vibrational spectrum, the levels are equally spaced. However, the degeneracies are peculiar to the five-dimensional oscillator.

The quadrupole moments and B(E2) values associated with the vibrational spectrum are also given in Fig. 9. These values are based on the following definition of the E2 operator (see, for instance, Kumar 1975 a for details of the calculation)

$$\mathcal{M}(E2,\mu) = Y\alpha_\mu , \quad (3.83a)$$

$$Y = 3(4\pi)^{-1} Z e R_0^2 , \quad (3.83b)$$

where α_μ is the deformation tensor in the lab-system, Z is nuclear charge, and R_0 is nuclear radius. This definition leads to the selection rule that only $\Delta N = 1$ transitions are allowed in the vibrational model. A special case of this selection rule leads to the general rule that the spectroscopic quadrupole moment vanishes for all states of the vibrational model. Note that this occurs not because the nucleus is spherical. This occurs because the model Hamiltonian is symmetric under the change of sign of deformation (prolate, oblate shapes occur with equal probability), while the quadrupole operator is antisymmetric.

The β-γ-dependent wave functions given above can be used to calculate the B(E2) values. For this purpose, one needs to transform the E2 operator of the lab system via the general transformation of a tensor of rank 2,

$$\mathcal{M}(E2,\mu) = \sum_\nu Q_\nu \mathcal{D}^2_{\mu\nu} (\phi,\theta,\psi), \quad (3.84a)$$

where Q_ν is the quadrupole tensor in the intrinsic system. With the definition of eqs. (3.83), one gets $Q_1 = Q_{-1} = 0$,

$$Q_0 = Y\beta\cos\gamma , \quad (3.84b)$$

and

$$Q_2 = Q_{-2} = \frac{1}{\sqrt{2}} Y\beta\sin\gamma . \quad (3.84c)$$

The matrix elements of the operators of eqs. (3.84) with respect to the basis wave functions of eqs. (3.42) (with $r_2 = r_3 = +1$) are given by

$$\langle I'M' | \mathcal{M}(E2,\mu) | IM \rangle = \sum_{KK'\nu} \langle \phi^{I'}_{M'K'} | \mathcal{D}^2_{\mu\nu} | \phi^I_{MK} \rangle$$

$$\times \langle A_{I'K'} | Q_\nu | A_{IK} \rangle$$

$$= (2I'+1)^{-\frac{1}{2}} C^{I\ 2\ I'}_{M\ \mu\ M'} \langle I' || \mathcal{M}(E2) || I \rangle, \quad (3.85a)$$

- 43 -

where the phase convention of Bohr and Mottelson (1969) for the reduced matrix element has been employed. Using the properties of the $\mathcal{D}$-functions, one finds (Kumar 1975 a).

$$M_{II'} = -\langle I' || \mathcal{M}(E2) || I \rangle$$

$$= -\sqrt{2I+1} \sum_{K \geq 0} \left[C^{I\ 2\ I'}_{K\ 0\ K} \langle A_{I'K} | Q_0 | A_{IK} \rangle \right.$$

$$+ \sqrt{\tfrac{1}{2}(1+\delta_{K0})} \left\{ C^{I\ 2\ I'}_{K\ 2\ K+2} \langle A_{I',K+2} | Q_{2'} | A_{IK} \rangle \right.$$

$$\left. \left. + C^{I\ 2\ I'}_{K+2\ -2\ K} \langle A_{I'K} | Q_{2'} | A_{I,K+2} \rangle \right\} \right], \qquad (3.85b)$$

where $Q_{2'} = \sqrt{2}\, Q_2$. The spectroscopic quadrupole moment is then given by

$$Q^s_I = \sqrt{\tfrac{16\pi}{5}} \langle I, M=I | \mathcal{M}(E2,0) | I, M=I \rangle$$

$$= -\sqrt{\tfrac{16\pi}{5}}\, C^{I\ 2\ I}_{I\ 0\ I}\, M_{II}\, (2I+1)^{-\tfrac{1}{2}}$$

$$= -\left[\frac{16\pi\, I(2I-1)}{5(I+1)(2I+1)(2I+3)} \right]^{\tfrac{1}{2}} M_{II}, \qquad (3.85c)$$

and the B(E2) values by

$$B(E2; I \to I') = (2I+1)^{-1}\, M^2_{II'}. \qquad (3.85d)$$

Note that the relations (3.84a, 3.85a-d) are not specific to the vibrational model, but are valid for the general collective problem (consistent with the symmetry $r_2 = r_3 = +1$). Using the vibrational model wave functions given above, one finds for the N=1, I=2 → N=0, I=0 transition:

$$M_{20} = -\sqrt{\tfrac{2}{5}}\, Y b \langle D^2 \rangle = -\sqrt{\tfrac{5}{2}}\, Y b, \qquad (3.86a)$$

$$B(E2; 2 \to 0) = \tfrac{1}{2} Y^2 b^2 = \tfrac{1}{5} Y^2 \beta^2_{rms}(0). \qquad (3.86b)$$

The above equation can be combined with that for the excitation energy of the first 2+ state,

$$E_2 = \hbar\omega = \sqrt{\tfrac{C}{B}}, \qquad (3.86c)$$

to relate the two model parameters to two observable quantities:

$$C = \frac{\hbar\omega}{b^2} = \frac{Y^2 E_2}{2B(E2;2\to 0)} \quad , \tag{3.86d}$$

$$\frac{B}{\hbar^2} = \frac{1}{\hbar\omega b^2} = \frac{Y^2}{2E_2 B(E2;2\to 0)} \quad . \tag{3.86e}$$

A table of the values of B,C derived in this way from the experimental values for the "spherical" nuclei, ranging from ^{16}O to ^{222}Rn, has been given by Wong (1968). The quantity C, which represents the softness of the nucleus against vibrations, shows maxima at closed shells and minima near the middle of the closed shells. In the oxygen region, the value of C drops from 2,454 MeV for ^{16}O to 55 MeV for ^{18}O. In the lead region, it drops from 1,437 MeV for ^{206}Pb to 14 MeV at ^{190}Os. The mass parameter, on the other hand, does not show quite so sharp maxima and minima. Its general A-dependence is given by the expression for the liquid drop of uniform density and irrotational flow:

$$\frac{B(LD)}{\hbar^2} = \frac{3M A R_0^2}{8\pi\hbar^2} = \frac{3(MC^2)AR_0^2}{8\pi(\hbar c)^2} = 4 \times 10^{-3} A^{5/3} \text{ MeV}^{-1}, \tag{3.86f}$$

where $R_0 = 1.2 A^{1/3}$ Fm has been used. The ratio B/B(LD) varies from 6 for ^{74}Se to 122 for ^{16}O.

A quantity of considerable interest is $\beta_{rms}(0)$ or the range parameter b, the two being simply proportional to each other in the present model. This quantity does not depend on the energy of the 2+ state. An analysis of the observed B(E2) values via the method of linear regression leads to the general empirical relation (Kumar 1979 a).

$$\beta_{rms} = \sqrt{\frac{5}{2}} b = \sqrt{\frac{5}{2}} (1.2 A^{-\frac{1}{2}}). \tag{3.86g}$$

Model B(E2) values connecting the low-energy states have been given previously (Kumar & Baranger 1967). But the signs of the E2 matrix elements, which are useful for some studies, were not given there. Both the signs and magnitudes are given in Table 5. The phase convention is based on the rotational model (see sect. III.D), and on the correspondence principle. The later is indicated in Fig. 9, which is a revised version of that proposed previously (Kumar 1975 a).

General properties of the vibrational model lead to the sum rule (Bohr & Mottelson 1975)

$$\sum_{I'} B(E2; N, I\to N-1, I') = N B(E2; 2\to 0), \tag{3.87a}$$

where I' represents different states allowed for (N-1) phonons. This rule has been employed in fig.10 to indicate the general trend of the B(E2) values, normalized to B(E2; 2→0). Note that in several instances, the summation in eq. (3.87a) reduces to a single term. For instance, in the case of the yrast states (I = 2N), this relations becomes

$$B(E2; I\to I-2) = \frac{1}{2} I B(E2; 2\to 0). \tag{3.87b}$$

Table 5. **Signs and Magnitudes of E2 Matrix Elements in the Vibrational Model.**

The magnitudes are in units of $\sqrt{B(E2; 2\rightarrow 0)}$. The matrix elements include the sign due to the i^2 factor, and obey the symmetry relation $M_{I'I} = (-1)^{I-I'} M_{II'}$.

N	N'	I	I'	$M_{II'}$	
1	0	2	0	$-\sqrt{5}$	a)
2	1	0 β	2	$-\sqrt{2}$	a)
		2 γ	2	$-\sqrt{10}$	a)
		4	2	$-\sqrt{18}$	a)
3	2	0 γγ	2 γ	$-\sqrt{3}$	a)
		2 β	0 β	$-\sqrt{7}$	a)
			2 γ	$-\sqrt{20/7}$	
			4	$-\sqrt{36/7}$	
		3 γ	2 γ	$\sqrt{15}$	a)
			4	$-\sqrt{6}$	
		4 γγ	2 γ	$-\sqrt{99/7}$	a)
			4	$-\sqrt{90/7}$	
		6	4	$-\sqrt{39}$	a)

a) Signs of these matrix elements are governed by the "rotational" phase convention employed here.

Thus, the B(E2) value connecting the yrast states increases linearly with angular momentum in the vibrational model.

D. Deformed (Rotational) Limit

The rotational model (Bohr & Mottelson 1953, 1975) is based on the assumption of rotation of a well-deformed, axially-symmetric nucleus. Small amplitude, harmonic vibrations around the equilibrium shape ($\beta_{eq} = \beta_0$, $\gamma_{eq} = 0°$) are also allowed. The corresponding collective Hamiltonian has the general form of eqs. (3.29-31) with (V is normalized to be zero at the equilibrium shape)

$$V = \frac{1}{2} C_{\beta\beta} (\beta - \beta_0)^2 + \frac{1}{2} C_{\gamma\gamma} \beta_0^2 \gamma^2, \tag{3.88a}$$

$$T_{vib} = \frac{1}{2} B_{\beta\beta} \dot{\beta}^2 + \frac{1}{2} B_{\gamma\gamma} \beta_0^2 \dot{\gamma}^2, \tag{3.88b}$$

and $\quad T_{rot} = \frac{1}{2} \mathcal{J}_0 (\omega_1^2 + \omega_2^2) + 2 B_{\gamma\gamma} \beta_0^2 \gamma^2 \omega_3^2. \tag{3.88c}$

The parabolic expansion of eq. (3.88a) violates the symmetry rquirement of eq. (3.10). Hence, it is valid only in the limit of extremely rigid vibrations ($C_{\beta\beta} \to \infty$, $C_{\gamma\gamma} \to \infty$) which would localize the collective wavefunction in a very small region around the equilibrium shape deformation. The vibrational kinetic energy of eq. (3.88b) involves the assumptions: $B_{\beta\gamma} = 0$, $B_{\beta\beta}$ and $B_{\gamma\gamma}$ are independent of deformation. The rotational kinetic energy of eq. (3.88c) is based on the assumptions: $\mathcal{J}_1 = \mathcal{J}_2 = \mathcal{J}_0 =$ constant, $\mathcal{J}_3 = 4 B_{\gamma\gamma} \beta_0^2 \gamma^2$ (see the general eq. 3.13 and the eq. 3.26 for the $\gamma \to 0°$ limit).

In spite of the rather drastic assumptions discussed above, the rotational model of Bohr & Mottelson is one of the most frequently employed models of nulcear physics. This is easily understood since "rotational" features are observed in practically all even-even nuclei and since the model is quite easy to use. However, it is rather unfortunate that enough care is not always exercised in separating truly experimental data from the model-dependent data. For instance, the rotational model is often employed to deduce the "experimental" deformations and moments of inertia. These quantitites are quite interesting. But their definitions are not unique (Bohr and Mottelson themselves give several definitions).

Returning to the rotational model described by eqs. (3.88a-c), we consider its solutions. The rotational part is quantized via eqs. (3.35) to give

$$\hat{T}_{rot} = \frac{\hbar^2}{2\mathcal{J}_0} (I_1^2 + I_2^2) + \frac{\hbar^2}{8 B_{\gamma\gamma} \beta_0^2 \gamma^2} I_3^2. \tag{3.89a}$$

The corresponding matrix elements are given by eqs. (3.59a-e), which become in the present case:

$$T_{KK} = \frac{\hbar^2}{2\mathcal{J}_0}[I(I+1) - K^2] + \frac{\hbar^2 K^2}{8B_{\gamma\gamma}\beta_0^2\gamma^2}, \qquad (3.89b)$$

$$T_{K,K+2} = T_{K+2,K} = 0. \qquad (3.89c)$$

Because of the assumption of axial symmetry, there is no K-mixing and K becomes a good quantum number. The diagonal matrix element, T_{KK}, contains two parts. The first part is a c-number, while the second part is an operator with respect to the γ degree of freedom. Hence, it is convenient to reorganize the collective Hamiltonian as follows:

$$H_{coll} = H_R + H_\beta + H_{\gamma\psi}, \qquad (3.90a)$$

$$H_R = \frac{1}{2}\mathcal{J}_0(\omega_1^2 + \omega_2^2), \qquad (3.90b)$$

$$H_\beta = \frac{1}{2}C_{\beta\beta}(\beta - \beta_0)^2 + \frac{1}{2}B_{\beta\beta}\dot{\beta}^2, \qquad (3.90c)$$

$$H_{\gamma\psi} = \frac{1}{2}C_{\gamma\gamma}\beta_0^2\gamma^2 + \frac{1}{2}B_{\gamma\gamma}\beta_0^2\dot{\gamma}^2 + 2B_{\gamma\gamma}\beta_0^2\gamma^2\dot{\psi}^2. \qquad (3.90d)$$

where we have employed the fact that although ω_3 is not simply equal to $\dot{\psi}$, I_3 depends only on $\dot{\psi}$ and, hence, only the ψ-dependent part is coupled to γ-vibrations.

Consider the solutions of $H_{\gamma\psi}$ of eq. (3.90c). First, we quantize the Hamiltonian via Pauli's method (see eq. 2.18) and write

$$\hat{H}_{\gamma\psi} = \frac{1}{2}C_{\gamma\gamma}\beta_0^2\gamma^2 - \frac{\hbar^2}{2B_{\gamma\gamma}\beta_0^2}\left[\frac{1}{\gamma}\frac{\partial}{\partial\gamma}\gamma\frac{\partial}{\partial\gamma} + \frac{1}{4\gamma^2}\frac{\partial^2}{\partial\psi^2}\right]. \qquad (3.91a)$$

Then, we make the transformation (compare with eqs. 3.73a-b)

$$\Gamma = \frac{\beta_0\gamma}{b}, \quad b = \sqrt{\frac{\hbar}{B_{\gamma\gamma}\omega_\gamma}}, \quad \omega_\gamma = \sqrt{\frac{C_{\gamma\gamma}}{B_{\gamma\gamma}}}, \qquad (3.91b)$$

and rewrite (3.91a) as

$$\hat{H}_{\gamma\psi} = \frac{1}{2}\hbar\omega_\gamma\left[\Gamma^2 - \frac{1}{\Gamma}\frac{\partial}{\partial\Gamma}\Gamma\frac{\partial}{\partial\Gamma} - \frac{1}{4\Gamma^2}\frac{\partial^2}{\partial\psi^2}\right]. \qquad (3.91c)$$

Write the trial wave function as

$$A(\Gamma,\psi) = f(\Gamma)\,g(\psi). \qquad (3.91d)$$

The requirement that A must be single-valued (and, hence, invariant under $\psi \to \psi + 2\pi$) leads to the solution

$$g(\psi) = \exp(iK\psi), \quad (K = \text{integer}) \tag{3.91e}$$

while $f(\Gamma)$ is a solution of the Schrödinger equation

$$\left(\Gamma^2 - \frac{1}{\Gamma}\frac{\partial}{\partial \Gamma}\Gamma\frac{\partial}{\partial \Gamma} + \frac{K^2}{4\Gamma^2}\right) f(\Gamma) = \eta f(\Gamma), \tag{3.91f}$$

where $\eta = E/(\frac{1}{2}\hbar\omega_\gamma)$. Now, we make a polynomial expansion of $f(\Gamma)$ and employ the requirement that $f(\Gamma) \to 0$ as $\Gamma \to \infty$, and write

$$f(\Gamma) = \sum_p a_p \Gamma^p \exp(-\frac{1}{2}\Gamma^2), \tag{3.92a}$$

where p is an integer. Substitution of (3.92a) into (3.91f) leads to the recursion relation

$$a_{p+2} = \frac{4(\eta-2p-2)}{K^2-(2p+4)^2} a_p \quad (K < 2p+4). \tag{3.92b}$$

Thus, the sum in (3.92a) terminates at $p = p_{max} = n_\gamma$ if

$$n = 2n_\gamma + 2, \text{ or } E = \hbar\omega_\gamma(n_\gamma + 1), \tag{3.92c}$$

and if $p > \frac{K}{2} - 2$. \hfill (3.92d)

Since the recursion relation connects only the $\Delta p = 2$ terms, we get the condition

$$p = \text{even (odd) for } n_\gamma = \text{even (odd)}. \tag{3.92e}$$

Thus, the function $f(\Gamma)$ is even (odd) in Γ if n_γ is even (odd). Hence, we have

$$f(-\Gamma) = (-1)^{n_\gamma} f(\Gamma). \tag{3.92f}$$

Allowed values of p and K are further restricted by the general requirement that the wave function be single-valued. Thus, the total wave function must be invariant under the 24 different ways of choosing $(\beta,\gamma,\phi,\theta,\psi)$ for a given set α_μ: there are six ways of labelling the three intrinsic axes (keeping them right-handed), and then there are four ways of orienting the axes. Bohr (1952) has shown that all these transformation can be obtained by repeated application of three of them: T_1 (which is the same as $R_1(\pi)$ of eq. (3.37), T_2 (rotation of $+90^0$ around the intrinsic-3-axis), T_3 (cicular permutation on the three intrinsic axes: $x'_1 = x_2, x'_2 = x_3, x'_3 = x_1$). Many consequences of these transformations have been taken into account in the treatments of sections III. B-C via the conditions based on the operators $R_1(\pi)$, $R_2(\pi)$, $R_3(\pi)$, see eqs. (3.42a-b). Others were taken into account by employing the basic tensors for each I value (see Table 3). However, these basic tensors are not employed in the rotational model. Hence, in order to find the allowed values of p,K, we need to consider one of the transformations mentioned above, namely T_2. The effect of this operator is to change (see Kumar & Baranger 1967)

$$\gamma \to -\gamma, \quad \psi \to \psi + \tfrac{1}{2}\pi .\qquad(3.93)$$

The condition of invariance under this transformation gives (using 3.91e and 3.92f):

$$(-1)^{K/2} (-1)^{n_\gamma} = 1. \qquad(3.93b)$$

The equations 3.92d-e, and 3.93b lead to the restrictions

$$p = \frac{K}{2}, \frac{K}{2}+2, \ldots, n_\gamma; \qquad(3.93c)$$

and $\quad K = \begin{cases} 0,4,\ldots,2n_\gamma & \text{if } n_\gamma = \text{even} \\ 2,6,\ldots,2n_\gamma & \text{if } n_\gamma = \text{odd} \end{cases}$. $\qquad(3.93d)$

According to eq. (3.93d), $K_{max} = 2n_\gamma$. This is analogous to the condition $I_{max} = 2N$ for the 5-D quadrupole vibrator (sect. III. C). Hence, in an analogous manner, one says that the γ-phonon carries a K value of two units.

Solutions of the one-dimensional β-vibration, governed by the Hamiltonian of eq. (3.90c), can be obtained in the same way. It is much simpler since there is no centrifugal barrier. There is no associated K-dependence. Hence, the β-vibrations are also called ΔK = 0-vibrations. The total wave function of the rotational Hamiltonian of eqs. (3.90a-d) is written as

$$|n_\beta n_\gamma IMK\rangle = f_{n_\beta}(\beta) f_{n_\gamma}(\gamma) \Phi^I_{MK}(\phi\theta\psi), \qquad(3.94)$$

where f_{n_β} is a solution of H_β (with $n_\beta = 0,1,2,\ldots$), f_{n_γ} is the γ-dependent part of the solution $H_{\gamma\psi}$ and is given by eqs. 3.92a, 3.93c-d; and Φ^I_{MK} is the symmetrized sum of the $\mathcal{D}$-functions (eqs. 3.42a-b with $r_2 = r_3 = \pm 1$). Note that the ψ-dependent part of the solution of $H_{\gamma\psi}$, given by eq. (3.91e), is identical to the ψ-dependent part of the $\mathcal{D}$-functions (Hence, Bohr & Mottelson call the motion governed by $H_{\gamma\psi}$ a γ-ψ vibration), and has been included in the "rotational wave-function" Φ^I_{MK}. Eigenvalues of the rotational model Hamiltonian are given by

$$W_{n_\beta n_\gamma KI} = \frac{\hbar^2}{2\mathcal{I}_0} [I(I+1) - K^2] + \hbar\omega_\beta(n_\beta + \tfrac{1}{2}) + \hbar\omega_\gamma(n_\gamma + 1), \qquad(3.95a)$$

where $\omega_\beta = \sqrt{\dfrac{C_{\beta\beta}}{B_{\beta\beta}}}, \quad \omega_\gamma = \sqrt{\dfrac{C_{\gamma\gamma}}{B_{\gamma\gamma}}}$. $\qquad(3.95b)$

The corresponding states can be considered as rotational bands built on different vibrational states. The lowest few bands are denoted by (see Fig.10):

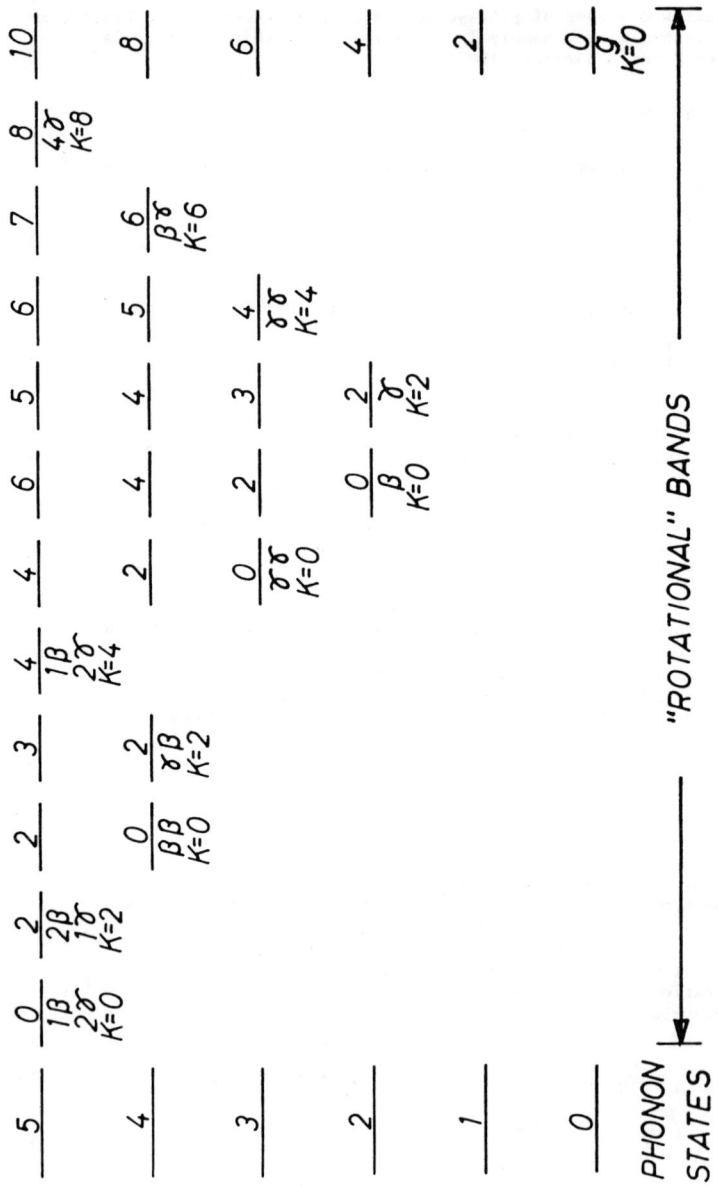

Fig. 10. Correspondence Principle for Phonon States of the Vibrational Model and Rotational Bands of the Rotational Model.

g-band: $n_\beta = 0$, $n_\gamma = 0$, $K = 0$; $I = 0, 2, 4, \ldots$ (3.96a)

β-band: $n_\beta = 1$, $n_\gamma = 0$, $K = 0$; $I = 0, 2, 4, \ldots$ (3.96b)

γ-band: $n_\beta = 0$, $n_\gamma = 1$, $K = 2$; $I = 2, 3, 4, \ldots$ (3.96c)

$\gamma\gamma_4$-band: $n_\beta = 0$, $n_\gamma = 2$, $K = 4$; $I = 4, 5, 6, \ldots$ (3.96d)

$\gamma\gamma_0$-band: $n_\beta = 0$, $n_\gamma = 2$, $K = 0$; $I = 0, 2, 4, \ldots$ (3.96e)

γβ-band: $n_\beta = 1$, $n_\gamma = 1$, $K = 2$; $I = 2, 3, 4, \ldots$ (3.96f)

ββ-band: $n_\beta = 2$, $n_\gamma = 0$, $K = 0$; $I = 0, 2, 4, \ldots$ (3.96g)

Since the only I-dependence of the rotational model wave function eq. 3.94) comes from Φ_{MK}^I, the relations for E2 moments (eqs. 3.85b-d) re simplified considerably. We can define an I-dependent "intrinsic" uadrupole moment connecting two bands α, α' (where α is a collective ndex for the band quantum numbers n_β, n_γ, K):

$$Q_{\alpha\alpha'}^{int} = \sqrt{\frac{16\pi}{5}} \langle \alpha' | q | \alpha \rangle \quad , \tag{3.97a}$$

$$q = \begin{cases} Q_0 & \text{if } \Delta K = |K-K'| = 0 \\ Q_2, & \text{if } \Delta K = 2, \ K \text{ or } K' = 0 \\ \frac{1}{\sqrt{2}} Q_2, & \text{if } \Delta K = 2, \ K \text{ and } K' \neq 0 \\ 0 & \text{otherwise} \end{cases} \tag{3.97b}$$

Some useful relations are:

$$M_{\alpha I, \alpha' I'} = - \sqrt{\frac{5(2I+1)}{16\pi}} \ C_{K \ K'-K \ K'}^{I \ 2 \ I'} \ Q_{\alpha\alpha'}^{int} \quad , \tag{3.98a}$$

$$B(E2; \alpha I \to \alpha' I') = \frac{5}{16\pi} \left(Q_{\alpha\alpha'}^{int}\right)^2 \left[C_{K \ K'-K \ K'}^{I \ 2 \ I'} \right]^2 , $$

$$= \frac{1}{16\pi} \left(Q_{\alpha\alpha'}^{int}\right)^2 \ (\text{if } K = K' = 0, \ I = 2, \ I' = 0) , \tag{3.98c}$$

$$Q_{\alpha I}^S = \frac{3K^2 - I(I+1)}{(I+1)(2I+3)} \ Q_{\alpha\alpha'}^{int} \quad , \tag{3.98d}$$

$$= -\frac{2}{7} Q_{\alpha\alpha}^{int} \ (\text{if } K=0, \ I=2), \tag{3.98e}$$

$$\frac{B(E2; \ \alpha(K=0)I \to \alpha'(K=0) \ I-2)}{B(E2: \ \alpha(K=0)2 \to \alpha'(K=0)0)} = \frac{15(I-1)I}{2(2I-1)(2I+1)} \xrightarrow[I\gg 1]{} \frac{15}{8}, \qquad (3.99a)$$

$$\frac{B(E2; \ \alpha(K=2)I \to \alpha'(K=2)I-2)}{B(E2; \ \alpha(K=2)4 \to \alpha'(K=2)2)} = \frac{63(I-3)(I-2)(I+1)(I+2)}{5(I-1)I(2I-1)(2I+1)} \xrightarrow[I\gg 1]{} \frac{63}{20}, \quad (3.99b)$$

$$\frac{B(E2; \ \alpha(K=2)I \to \alpha'(K=2)I-1)}{B(E2; \ \alpha(K=2)4 \to \alpha'(K=2)2)} = \frac{504(I-2)(I+2)}{5(I-1)I(I+1)(2I+1)} \xrightarrow[I\gg 1]{} 0, \qquad (3.99c)$$

A complete set of formulae for the E2 matrix elements allowed in the rotational model are given in Table 6. These can be combined with eqs. (3.85c-d) to obtain the quadrupole moments and the B(E2) values. The low-energy spectrum, B(E2) values, and quadrupole moments are indicated in Fig. 4.

The sign of the E2 matrix elements plays an important role in the deduction of B(E2) values and the quadrupole moments from the Coulomb Excitation probabilities (Alder and Winther 1966), as well as in the E2/M1 mixing ratios (Kumar 1975a). Since the sign of a wave function is completely arbitrary, a sign convention based on the rotational model is often employed. Hence, the signs and magnitudes of some of the matrix elements employed for this purpose are listed in Table 7. This table does note give the matrix elements for the band-heads. Their phases can be fixed by considering the most common mode of decay for such states — for example, $0_\beta \to 2_g$, $2_\gamma \to 2_g$, $0_{\gamma\gamma} \to 2_\gamma$, $4_{\gamma\gamma} \to 2_\gamma$, $2_{\gamma\beta} \to 2_\beta$, $0_{\beta\beta} \to 2_\beta$. The sign of the corresponding transition matrix elements can be obtained from Table 6. It is negative for all the mentioned transitions.

The rotational model discussed above depends on six parameters: B_0, $\mathcal{J}_0$, $B_{\gamma\gamma}$, $C_{\gamma\gamma}$, $B_{\beta\beta}$, $C_{\beta\beta}$. These parameters can be determined from six observable quantities — three energies and three B(E2) values. The three energy relations come from eqs. (3.95a-b):

$$\frac{\mathcal{J}_0}{\hbar^2} = \frac{3}{E_{2_g}} , \ E_{2_g} = W_{0002} - W_{0000}, \qquad (3.100a)$$

$$\hbar\omega_\beta = \hbar\sqrt{\frac{C_{\beta\beta}}{B_{\beta\beta}}} = E_{0_\beta} = W_{1000} - W_{0000}, \qquad (3.100b)$$

$$\hbar\omega_\gamma = \hbar\sqrt{\frac{C_{\gamma\gamma}}{B_{\gamma\gamma}}} = E_{2_\gamma} - \frac{1}{3} E_{2_g}, \ E_{2_\gamma} = W_{0122} - W_{0000}. \qquad (3.100c)$$

In order to relate the B(E2) values to the model parameters, we need to calculate the "intrinsic" moments defined by eqs. (3.97a-b). For this purpose, we note that the quadrupole operator is related to the deformation parameters by eqs. (3.83-84). These equations are rewritten as

Table 6. Signs and Magnitudes of E2 Matrix Elements in the Rotational Model. The index α represents a rotational band (n_β, n_γ, K). See the caption to Table 5 for the phase convention, and the symmetry realtion of $M_{\alpha I, \alpha' I'}$.

K'	I'	$M_{\alpha I, \alpha' I'} / \left[Q^{int}_{\alpha\alpha'} / \sqrt{16\pi} \right]$
K	I	$\left[I(I+1) - 3K^2 \right] \sqrt{\dfrac{5(2I+1)}{I(I+1)(2I-1)(2I+3)}}$
	I-2	$- \sqrt{\dfrac{15(I-K-1)(I-K)(I+K-1)(I+K)}{2(I-1)I(2I-1)}}$
	I-1	$K \sqrt{\dfrac{15(I-K)(I+K)}{(I-1)I(I+1)}}$
K±2	I	$- \sqrt{\dfrac{15(2I+1)(I\mp K-1)(I\mp K)(I\pm K+1)(I\pm K+2)}{2I(I+1)(2I-1)(2I+3)}}$
	I-2	$- \sqrt{\dfrac{5(I\mp K-3)(I\mp K-2)(I\mp K-1)(I\mp K)}{4(I-1)I(2I-1)}}$
	I-1	$\pm \sqrt{\dfrac{5(I\mp K-2)(I\mp K-1)(I\mp K)(I\pm K+1)}{2(I-1)I(I+1)}}$

Table 7. Signs and Magnitudes of Some E2 Matrix Elements Utilized for the "Rotational" Phase Convention. The matrix elements are normalized according to $\sqrt{B(E2\ ;\ I=2,\ K=0 \rightarrow I=0,\ K=0)} = 1$. Note that M_{fi} equals $(-1)^{I_i - I_f} M_{if}$.

I_i	I_f	M_{if}	I_i	I_f	M_{if}	I_i	I_f	M_{if}
\multicolumn{3}{c}{$K=0 \rightarrow K=0$}			$K=2 \rightarrow K=2$			$K=4 \rightarrow K=4$
2	0	-2.236	3	2	3.536	5	4	4.243
4	2	-3.586	4	2	-2.315	6	4	-2.023
6	4	-4.523	5	3	-3.240	7	5	-3.011
8	6	-5.292	6	4	-3.908	8	6	-3.761
10	8	-5.960	7	5	-4.475	9	7	-4.372
12	10	-6.561	8	6	-4.910	10	8	-4.894
14	12	-7.110	9	7	-5.321	11	9	-5.351
16	14	-7.620	10	8	-5.694	12	10	-5.762
18	16	-8.098	11	9	-6.039	13	11	-6.137
20	18	-8.549	12	10	-6.361	14	12	-6.483
22	20	-8.977	13	11	-6.665	15	13	-6.807
24	22	-9.385	14	12	-6.954	16	14	-7.111
26	24	-9.777	15	13	-7.229	17	15	-7.399
28	26	-10.153	16	14	-7.493	18	16	-7.673
30	28	-10.516	17	15	-7.747	19	17	-7.936
32	30	-10.867	18	16	-7.992	20	18	-8.188
34	32	-11.207	19	17	-8.229	21	19	-8.431
36	34	-11.537	20	18	-8.458	22	20	-8.665

$$Q_0 = Y \beta \cos\gamma \approx Y[\beta_0 + (\beta - \beta_0)] \,, \qquad (3.101a)$$

$$Q_{2'} = Y \beta \sin\gamma \approx Y\beta_0 \gamma. \qquad (3.101b)$$

Using the facts that β_0 is a constant, that $(\beta - \beta_0)$ is the coordinate of one-dimensional β-vibrations, and that γ is the "radial" coordinate of the two-dimensional γ-ψ-vibrations, one gets

$$<\alpha'|Q_0|\alpha> = Y\beta_0 \quad (\text{if } \alpha' = \alpha), \qquad (3.101c)$$

$$= Y \left(\frac{\hbar n_{\beta>}}{2B_{\beta\beta}\omega_\beta}\right)^{\frac{1}{2}} \quad (\text{if } \Delta n_\beta = 1, \Delta n_\gamma = 0), \qquad (3.101d)$$

$$<\alpha'|Q_{2'}|\alpha> = Y \left(\frac{\hbar n_{\gamma>}}{B_{\gamma\gamma}\omega_\gamma}\right)^{\frac{1}{2}} \quad (\text{if } \Delta n_\gamma = 1, \Delta n_\beta = 0).$$

The three basic B(E2) values for the rotational model are:

$$B(E2; 2_g \to 0_g) = \frac{1}{16\pi}\left(Q_{gg}^{int}\right)^2 = \frac{1}{5} Y^2 \beta_0^2 \,, \qquad (3.102a)$$

$$B(E2; 2_\beta \to 0_g) = \frac{1}{16\pi}\left(Q_{\beta g}^{int}\right)^2 = \frac{\hbar^2 Y^2}{10 B_{\beta\beta}(\hbar\omega_\beta)} \,, \qquad (3.102b)$$

and $\quad B(E2; 2_\gamma \to 0_g) = \frac{1}{16\pi}\left(Q_{\gamma g}^{int}\right)^2 = \frac{\hbar^2 Y^2}{5 B_{\gamma\gamma}(\hbar\omega_\gamma)} \,. \qquad (3.102c)$

The six observable quantities mentioned above determine the model parameters completely. These can be combined with the relations for the energies, B(E2) values, and the quadrupole moments to predict a wealth of observable data. In fact, many ratios of these quantities are independent of the model parameters.

In view of the simplicity of the rotational model, it is not surprising that the observed energies and E2 moments show deviations from the model - which are sometimes quite substantial (e.g. Hamilton 1972). What is surprising is that many of the rotational features are so universal in atomic nuclei. One of the most important of these is the Alaga rule (Alder et al. 1956) that the B(E2) values are proportional to the squares of certain Clebsch-Gordon coefficients (see eq. 3.89b). This rule appears to have some validity even at quite high spins where the Coriolis force becomes so strong that an increase in nuclear spin is accompanied by a <u>decrease</u> in rotational frequency - the phenomenon called Backbending (Lieder & Ryde 1978).

On the other hand, the simple "rotational" relations must and do break down at high enough spins. The estimated spin is of the order of 100 in heavy nuclei, but it is only 4 (^{8}Be) - 8 (^{20}Ne) in light nuclei where the B(E2) value vanishes (Bohr & Mottelson 1975).

IV. Deformed Single Particle Motion

A. Rainwater-Nilsson Hamiltonian and its Parameters

One of the most amazing aspects of nuclear physics is the validity of the nuclear shell model, especially as represented by the simple oscillator potential, in spite of the many complications of the nucleon-nucleaon interaction and of the nuclear many-body problem. A basic reason for this validity was given by M. Mayer (1950), as already mentioned in sect. I.A. This reasoning was made for the spherical shell model. But it is equally valid for the deformed shell model, which is a generalization of the spherical shell model.

Rainwater (1950) made this generalization in a beautifully simple way. The spherical oscillator potential, $\frac{1}{2}M\omega^2 r^2$, is modified to $\frac{1}{2}M[\omega_1^2(x^2+y^2) +\omega_z^2 z^2]$ if the nucleus is deformed and axially symmetric along the z-axis. This potential was combined with the usual ℓ^2- and $\vec{\ell}\cdot\vec{s}$-terms of the spherical shell model by Nilsson (1955), who presented widely employed graphs of deformed-single-particle-level energies as functions of a nuclear deformation parameter.

This model was further generalized to asymmetric shapes ($\omega_x \neq \omega_y \neq \omega_z$) by Newton (1960), who presented some quite useful tables of level energies and wave functions.

We discuss below a more recent version of the Rainwater-Nilsson Hamiltonian and its parameters (Kumar et al. 1977). The Hamiltonian is written as

$$H_{av} = \frac{\vec{p}^2}{2M} + \tfrac{1}{2}M(\omega_1^2 x_1^2 + \omega_2^2 x_2^2 + \omega_3^2 x_3^2)$$
$$+ \hbar\omega_0 [v_{\ell\ell}(\vec{\ell}^2 - \langle\vec{\ell}^2\rangle_N) + v_{\ell s} \vec{\ell}\cdot\vec{s}], \qquad (4.1a)$$

where $\langle\vec{\ell}^2\rangle_N = \tfrac{1}{2} N(N+3)$, (4.1b)

N is the major oscillator shell number (0,1,2,...), M is nucleon mass, $\vec{p}$ ($\vec{\ell}$) its linear (angular) momentum, $\vec{s}$ its spin, x_k (k=1,2,3) its position coordinate, ω_k is oscillator frquency along the intrinsic-axis k, $\hbar\omega_0$ is the average oscillator frequency defined via the relation

$$\omega_1\omega_2\omega_3 = \omega_0^3, \qquad (4.1c)$$

and $v_{\ell\ell}$, $v_{\ell s}$ are the strength constants for $\vec{\ell}\cdot\vec{\ell}$, $\vec{\ell}\cdot\vec{s}$ interactions. The term $<\vec{\ell}^2>_N$ is included so as to keep the average of the single-particle energies the same as in the absence of the $\vec{\ell}^2$ term (Gustafson et al. 1967).

The Hamiltonian of eqs. (4.1a-c) depends on five parameters: $\omega_1, \omega_2, \omega_3$, $v_{\ell\ell}, v_{\ell s}$. The first three of these are expressed in terms of the average value ω_0 (see eq. 4.1c) and two nuclear shape parameters. In order to make the later connection, one first relates the oscillator frquencies to nuclear radii (semi-axis lengths). The argument is that since the nuclear force has a short range, the equipotential surfaces will have the same shape as the nucleus (one really means the equivalent ellipsoid which has the same moments $<x_k^2>$ as the actual nucleus with a diffuse surface.). Hence, the oscillator frequencies ω_k are inversely proportional to nuclear radii:

$$\omega_k = \omega_0 R_0 / R_k \quad (k=1,2,3). \quad (4.2)$$

The three nuclear radii are related to the average nuclear radius R_0 via the volume conservation condition,

$$R_1 R_2 R_3 = R_0^3 , \quad (4.3)$$

and to two nuclear shape variables (β, γ) via

$$R_k = R_0 \exp[\delta \cos(\gamma - \frac{2}{3}\pi k)] , \quad (4.4)$$

$$\delta = \beta \sqrt{\frac{5}{4\pi}} . \quad (4.5)$$

Equation (4.4) represents the Hill-Wheeler (1953) definition of nuclear deformation. Equation (4.5) relates their deformation parameter δ to Bohr's (1952) deformation parameter β. In fact, the two definitions are equivalent only to first order in deformation, since the Bohr definition is equivalent to expanding the right-hand side of eq. (4.4) and keeping only the first order term in δ. In fact, many other definitions of deformation exist in the literature (e.g. Bohr & Mottelson 1975). The advantage of the Hill-Wheeler definition is that the volume conservation condition (4.3) is satisfied exactly to all orders in δ.

Returning to our discussion of the model parameters, we note that the three oscillator frequencies are related to ω_0, $\beta(\delta)$, γ via

$$\omega_k = \omega_0 \exp[-\delta \cos(\gamma - \frac{2}{3}\pi k)] . \quad (4.6)$$

Of these, the average frequency ω_0 is related to the nuclear radius (that is the radius of the sphere having the same volume as the nucleus) via the standard oscillator relation

$$<r^2> = \frac{3}{5} R_0^2 = \sum_{i=1}^{A} (N_i + 3/2) \frac{\hbar}{M\omega_0} . \quad (4.7)$$

The observations that $R_0 \sim (1.2 \text{ fm}) A^{1/3}$ and that $<r^2>_p \approx <r^2>_n$ are combined with eq. (4.7) in the Bohr-Mottelson relations

$$\begin{pmatrix} \hbar\omega_{0p} \\ \hbar\omega_{0n} \end{pmatrix} = \frac{41 \text{ MeV}}{A^{1/3}} \begin{pmatrix} 1 - \frac{N-Z}{3A} \\ 1 + \frac{N-Z}{3A} \end{pmatrix} . \tag{4.8}$$

The corresonding isospin-dependent oscillator length constants are given by

$$b_\tau^2 = \frac{\hbar}{M\omega_{0\tau}} \quad . \quad (\tau = p,n) \tag{4.9}$$

As regards the shape parameters β and γ, they are not really model parameters in the Dynamic Deformation Theory. Instead, they are dynamic variables in terms of which the collective Schrödinger equation (discussed in ch. II, III) is expressed. The limits of integration are $\beta = 0.0 - 0.8$, $\gamma = 0° - 60°$. The upper limit of β is based on the observation that the β-values deduced from the experimental B(E2) values (Stelson & Grodzins 1965) are 0.2 - 0.4 for well-deformed nuclei and on the experience with the β_{max} value needed for the convergence of the collective solutions. The upper limit of γ is based on the symmetry properties of eq. (4.4) according to which $\gamma \to \gamma \pm 120°$ and $\gamma \to \pm \gamma$ are equivalent to a relabelling of the intrinsic axes.

The remaining two parameters of the model, $v_{\ell s}$ and $v_{\ell\ell}$, are usually chosen from an analysis of the spectra of odd-A nuclei. Extensive studies of this type were made by Nilsson et al. (Nilsson 1955, Gustafson et al. 1967). It was found that two parameters were not adequate for all nuclei. Hence, different sets of parameters were employed for different regions, and sometimes even for each and every nucleus. In order to avoid this problem, and also to avoid the necessity of making quite lengthy calculations for each nucleus, Kumar et al. (1977) adopted the following procedure for the determination of $v_{\ell s}$, $v_{\ell\ell}$: Starting with the highest magic number (258) for the configuration space of N=0-8 shells and moving down to lower magic numbers (184, 126, 82, 50, 20), the Strutinsky shell correction (see sect. VI.A) was calculated as a function of ($v_{\ell s}$, $v_{\ell\ell}$). The set of two values, which minimized the shell correction for that magic number, was adopted for the shell above (N=8 for the case of magic number = 258). This procedure is a variation of that employed by Yariv et al. (1976). This procedure was repeated for each shell with N=8,7,...,3. Values of $v_{\ell s}$, $v_{\ell\ell}$ were obtained for N=1,2 by fitting the spectra of magic number ±1 nuclei. The values obtained in this way were given previously (Kumar et al. 1977, Table I).

It may seem strange to adopt different $v_{\ell s}$, $v_{\ell\ell}$ values for different oscillator shells. But one has to keep in mind that this is just a way of parameterizing the spherical single- particle energies. These energies are employed as fitting parameters in most shell model based calculations. Furthermore, the $\vec{\ell}\cdot\vec{s}$ and $\vec{\ell}^2$ terms affect only the spherical level energies. They are diagonal with respect to the spherical oscillator wave functions. Hence, the basis wave functions are unaffected by the local variation of $v_{\ell\ell}$, $v_{\ell s}$ strengths from shell to shell.

The great advantage of the present procedure of choosing the $v_{\ell\ell}$, $v_{\ell s}$ is that the DSPL basis can be made independent of Z and A. How this is achieved is the topic for the following section.

B. Scaling Method Leading to a Common Basis for All Nuclei

A great advantage of the oscillator potential is that all energies and lenths can be scaled in terms of two parameters - $\hbar\omega_{0p}$, $\hbar\omega_{0n}$. This advantage was exploited to a certain extent in the early versions of the Nilsson model - where Nilsson diagrams were presented for different nuclear regions - but it seems to have been discarded in most of the current versions of the model where different $v_{\ell\ell}$, $v_{\ell s}$ values are adopted for each nucleus. The parameterization discussed above allows us to take full advantage of the scaling properties of the oscillator model.

We make the usual transformation

$$r \to \rho = r/b, \quad x_k \to \rho_k = x_k/b, \qquad (4.10)$$

where b is the oscillator lenth parameter related to ω_0 via

$$b^2 = \frac{\hbar}{M\omega_0} = \frac{\hbar^2 c^2}{Mc^2(\hbar\omega_0)} . \qquad (4.11)$$

Then, the Rainwater-Nilsson Hamiltonian (4.1) can be rewritten as (utilizing eq. 4.6)

$$H_{av} = \hbar\omega_0 \, h_{av}, \qquad (4.12)$$

$$h_{av} = h_{sph} + h_{def}, \qquad (4.13)$$

$$h_{sph} = -\tfrac{1}{2}\nabla_\rho^2 + \tfrac{1}{2}\rho^2 + v_{\ell\ell}(\vec{\ell}^2 - \langle\vec{\ell}^2\rangle_N) + v_{\ell s}\,\vec{\ell}\cdot\vec{s}, \qquad (4.14)$$

$$h_{def} = \beta_{00}\,\rho^2 - \beta_0^P\,Q'_{20} - \beta_2^P\,Q'_{22'}, \qquad (4.15)$$

where $\nabla_\rho^2 = \sum_k \dfrac{\partial^2}{\partial \rho_k^2}$, (4.16)

$$Q'_{20} = Q_{20}/b^2 = \rho^2 Y_{20} = \sqrt{\frac{5}{16\pi}}\,(2\rho_3^2 - \rho_1^2 - \rho_2^2), \qquad (4.17)$$

$$Q'_{22'} = Q_{22'}/b^2 = \frac{1}{\sqrt{2}}\,\rho^2(Y_{22} + Y_{2,-2}) = \sqrt{\frac{15}{16\pi}}\,(\rho_1^2 - \rho_2^2), \qquad (4.18)$$

$$\beta_{00} = \frac{1}{6} [\exp(-2\delta_x) + 2\exp(\delta_x)\cosh\delta_y - 3] , \qquad (4.19)$$

$$\beta_0^P = \frac{1}{6}\sqrt{\frac{16\pi}{5}} [\exp(\delta_x)\cosh\delta_y - \exp(-2\delta_x)] , \qquad (4.20)$$

$$\beta_{2'}^P = \frac{1}{6}\sqrt{\frac{48\pi}{5}} \exp(\delta_x)\sinh\delta_y , \qquad (4.21)$$

and $\quad \delta_x = \delta\cos\gamma, \; \delta_y = \sqrt{3}\,\delta\sin\gamma$. $\qquad (4.22)$

The single-particle Schrödinger equation for the Hamiltonian of eq. (4.13) is

$$h_{av} |p\rangle = \epsilon_p |p\rangle . \qquad (4.23)$$

Its eigenvalue ϵ_p and eigenvectors $|p\rangle$ are determined by expanding $|p\rangle$ in a spherical basis:

$$|p\rangle = \sum_a C_{ap} |a\rangle , \qquad (4.24)$$

where "a" is a collective index for the four quantum numbers $(N\ell j\Omega)$. Since the Hamiltonian is time-reversal-invariant, the Hamiltonian-Matrix $h_{ab} = \langle b|h_{av}|a\rangle$ breaks up into two unconnected blocks: one consisting of states with

$$\Omega = \frac{1}{2}, -\frac{3}{2}, \frac{5}{2}, -\frac{7}{2}, \ldots (-1)^{j-\frac{1}{2}} j , \qquad (4.25)$$

and the other consisting of states with $\Omega = -\frac{1}{2}, \frac{3}{2}, \ldots$

The deforming part of the Hamiltonian, h_{def} of eq. (4.15), gives mixings of type $\Delta N = \pm 2$, $\Delta\ell = \pm 2$, $\Delta j = \pm 1$ or ± 2, $\Delta\Omega = \pm 2$. Hence, all four quantum numbers of the spherical basis are destroyed. However, parity and isospin remain good quantum numbers. Also, the time-reversal-conjugate states $|p\rangle$ and

$$|\bar{p}\rangle = -\tau|p\rangle \qquad (4.26$$

are degenerate $(\epsilon_{\bar{p}} = \epsilon_p)$. Hence, the total configuration space of $N = 0-8$ shells, which can accomodate 660 neutrons and protons, yields two irreducible matrices of size 95×95 and 70×70.

The mixing coefficients C_{ap} are independent of the scaling factors ($\hbar\omega_0$ for the Hamiltonian, or b for the lengths) and, hence, of Z and A. The total wave function depends on Z and A via the scaling factors since, for instance, the range of the Gaussian is unity in the ρ-space but is b in the r-space. However, for most practical purposes, one does not need to know the r-dependence of the wave function. All observables calculated in the ρ-space can be directly converted to those in the r-space via the scaling factors. Thus, all energies are multiplied by $\hbar\omega_0$ (actually $\hbar\omega_{0\tau}$, but we have dropped the subscript τ in order to simplify the notation), and the eigenvalues of H_{av} are given by

$$\eta_p = \hbar\omega_0 \, \epsilon_p. \tag{4.27}$$

All lengths are multiplied by the length scaling factor $b = \sqrt{\hbar/(M\omega_0)}$.

Thus, it is possible to work in a large configuration space of 660 particles so that all nucleons can be treated as active nucleons. This is the great advantage over the pairing-plus-quadrupole model (Kumar 1975a and previous references cited there) - where it was necessary to work in a truncated space of two major shells (equivalent to 170 particles) and, hence, to employ effective charges and inertial renormalization factors. All such "inert core" parameters are eliminated in the present version of the Dynamic Deformation Theory.

The configuration space employed in some other versions of the Nilsson model is also as large as the one employed here. However, the $\Delta N = 2$ mixing is neglected. This mixing, which turns out to be rather important for obtaining the correct mass parameters and moments of inertia, is included here. This leads to a substantial increase in the computation time. One needs to diagonalize a $95*95$ and a $70*70$ matrix for each of 92 mesh points in the β-γ plane (discussed in ch. III) and then to calculate the expectation values of nine different operators for each case. The calculation of the corresponding 6 million matrix elements required 8 hours of UNIVAC at Orsay (July 1976). However, the same matrix elements (stored on tapes) have been employed since then for different even-even nuclei ranging from ^{12}C to ^{240}Pu.

Thus, the scaling method discussed above has led to a common basis for all nuclei. Naturally, the basis is not the best one for each and every nucleus. But it has made it possible to make a global study of practically all nuclei.

An example of some of the calculated single-particle levels is given in Fig.11. Note that these levels are the same for protons and neutrons (except for the scaling factors) in the present theory. Such levels are quite useful for getting a preliminary idea about the shape-dependence of a particular nucleus. The circled numbers indicate the approximate location of the Fermi-energy for that number of nucleons and for that particular shape. The rich diversity of the single-particle level density near the Fermi surface leads to the rich diversity in the average shapes of the total (collective) nuclear states.

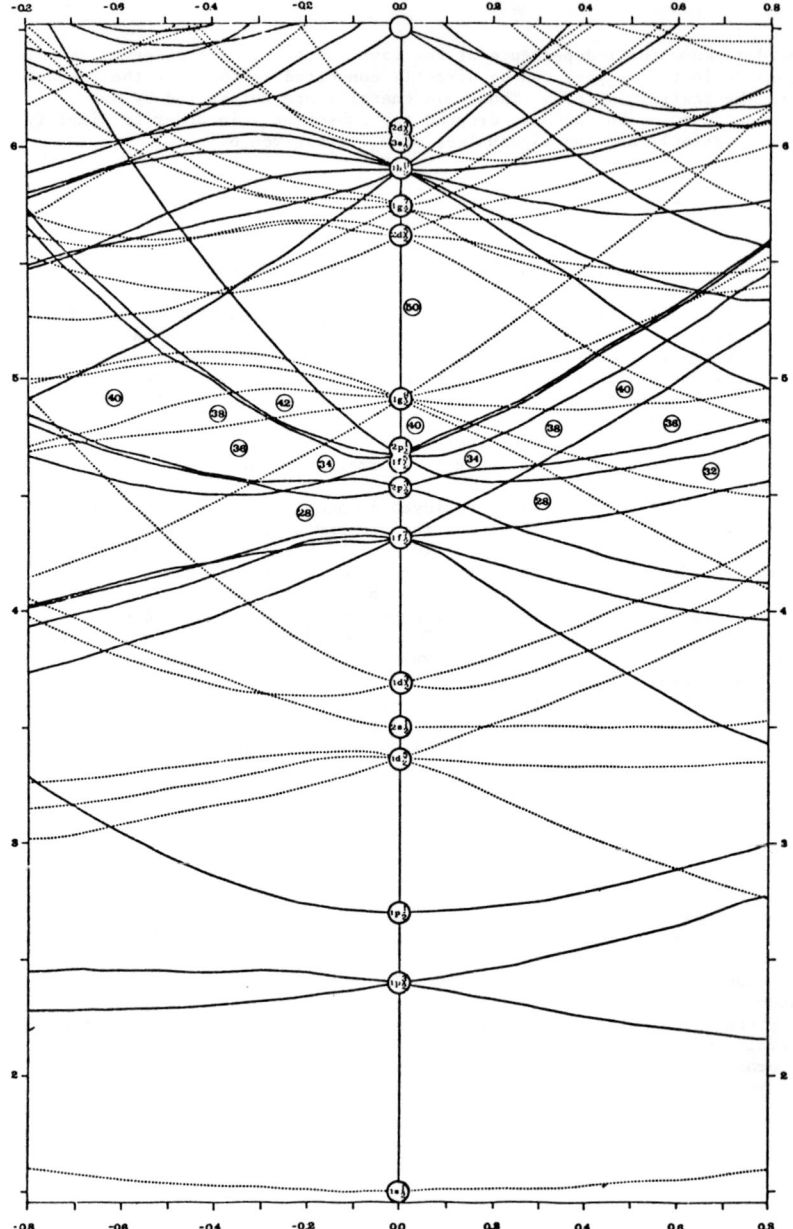

Fig.11. Partial Level Scheme of the Deformed Single-Particle Basis. The circled numbers (28, 40, ...) represent the approximate location of the Fermi level for that number of protons or neutrons. The level energies, in units of $\hbar\omega$, are given as functions of β.

V. Pairing Correlations

A. Improved BCS Theory Including The Particle-Hole Channel

The shell-model philosophy can be expressed as follows. One writes the total nuclear Hamiltonian as a sum of kinetic energy and interaction energy,

$$H = T + V_{int} . \qquad (5.1a)$$

One adds and substracts a term U, which represents the average-one-body potential created by the many-body nucleon-nucleon interaction V_{int}:

$$H = (T + U) + (V_{int} - U)$$

$$= H_{av} + V_{res}$$

$$= \sum_{pq} H_{pq} c_p^+ c_q + \frac{1}{4} \sum_{pqrs} V_{pqrs} c_q^+ c_p^+ c_r c_s , \qquad (5.1b)$$

where c_p^+ (c_q^+) is a creation (destruction) operator for the single-particle state $|p>$,

$$H_{pq} = <p | H_{av} | q> , \qquad (5.1c)$$

and

$$V_{pqrs} = <pq|V_{res}|rs> - <qp | V_{res} | rs> . \qquad (5.1d)$$

Then, one assumes that the major effects of the nucleon-nucleon interaction are already included in H_{av}. Hence, one can employ a comparatively simple form for the residual interaction, V_{res}.

In the spherical-shell model, one assumes that the average potential U is spherically symmetric. Then, a single-particle state $|p>$ is charachterized by four quantum numbers: n_p, ℓ_p, j_p, Ω_p, where Ω_p is the projection of $\vec{j}_p$ on the intrinsic-z-axis.

In the deformed-shell-model, one attempts to include even more of the nucleon-nucleon interaction into U by allowing it to be deformed (or spherical, depending on the dynamics of the nucleons). An example of such a deformed-shell model has been discussed above (ch. IV). In this case, one argues that so much of the interaction is already included in H_{av} that a particularly simple form of V_{res} is sufficient for the low-energy properties, namely that coming from the pairing force of the type employed by Bardeen, Cooper, and Schrieffer (1957) for explaining the phenomenon of superconductivity.

In a spherical basis, the pairing force acts with a constant strength between any two pairs of nucleons coupled to $J = 0$. In a deformed basis, the same force leads to the matrix elements

$$V_{pqrs} = - G s_p s_r \delta_{q\bar{p}} \delta_{s\bar{r}} , \qquad (5.2a)$$

where p,q,r,s represent different solutions of H_{av}, i.e.

$$H_{pq} = \eta_p \delta_{pq} , \qquad (5.2b)$$

$\bar{p}$ represents the time-reversal-conjugate state of p, s_p is the associated phase factor defined such that

$$s_p^2 = 1, \quad s_{\bar{p}} = -s_p, \quad s_p \xrightarrow[\beta \to 0]{} (-1)^{j_p - \ell_p + \Omega_p}, \qquad (5.2c)$$

and G is the strength of the pairing force.

The pairing force is usually treated by employing the Bogolyubov transformation from the particle, hole states to the quasi-particle states defined by

$$a_p^+ = u_p c_p^+ - v_p s_p c_{\bar{p}} , \qquad (5.3a)$$

$$a_p = u_p c_p - v_p s_p c_{\bar{p}}^+ , \qquad (5.3b)$$

where u_p, v_p are the mixing amplitudes with the normalization condition

$$u_p^2 + v_p^2 = 1 . \qquad (5.3c)$$

The quasi-particle operators of (5.3a-c) do not conserve the particle number. Hence, one works with the constrained Hamiltonian

$$H' = H - \lambda \hat{N} , \qquad (5.4)$$

where H is given by (5.1b), λ is the Lagrange multiplier (Fermi energy), and $\hat{N}$ is the particle-number operator

$$\hat{N} = \sum_p c_p^+ c_p . \qquad (5.5)$$

The Lagrange multiplier is determined via the constraint condition

$$\langle \hat{N} \rangle = N , \qquad (5.6)$$

where N is the number of nucleons in the system under consideration.

The u,v factors are determined by employing the Bogolyubov conditions:

<u>Condition I.</u> Terms connecting 0,2 quasi-particle states must vanish, i.e.

$$\langle pq | H' | 0 \rangle = 0 \quad \text{for all p,q} . \qquad (5.7a)$$

<u>Condition II.</u> The one quasi-particle terms are diagonal, i.e.

$$\langle q | H' | p \rangle = E_p \delta_{pq} . \qquad (5.7b)$$

The zero, one, two quasi-particle states are defined by the relations

$$a_p |0\rangle = 0 \quad \text{for all } p , \tag{5.8a}$$

$$|p\rangle = a_p^+ |0\rangle , \tag{5.8b}$$

and $\quad |pq\rangle = a_p^+ a_q^+ |0\rangle . \tag{5.8c}$

While the condition I leads to an approximate diagonalization of the Hamiltonian for even-even nuclei, the condition II yields a simultaneous diagonalization of H' for odd-A nuclei. Also, this procedure allows for the inclusion of a large configuration-space, as compared to a direct matrix-diagonalization where the computation time increases much more rapidly with the size of the configuration-space. Thus, the Bogolyubov quasi-particle transformation, supplemented by the conditions (5.7a,b), provides a rather powerful tool for the nuclear many-body problem.

When the quasi-particle transformation and the Bogolyubov conditions are applied to a general, two-body-interaction Hamiltonian such as that of eq. (5.1a,b), one finds that eqs. (5.7a,b) become on using the standard anticommutation relations for fermions (see for instance, Kumar 1975 b)

$$(u_p v_q + v_p u_q)(H_{pq} - \lambda \delta_{pq} + A_{pq}) - (u_p u_q - v_p v_q) B_{pq} = 0, \tag{5.9a}$$

$$(u_p u_q - v_p v_q)(H_{pq} - \lambda \delta_{pq} + A_{pq}) + (u_p v_q + v_p u_q) B_{pq} = E_p \delta_{pq} , \tag{5.9b}$$

where $A_{pq} = \sum_c' v_c^2 (V_{pcqc} + V_{pc\bar{q}\bar{c}}) , \tag{5.9c}$

$$B_{pq} = \sum_c' u_c v_c s_q s_c V_{p\bar{q}c\bar{c}} , \tag{5.9d}$$

and $\sum'$ represents summation over half the total number of single-particle states (for instance, those defined by eq. 4.25).

The matrix A_{pq} of eq. (5.9c) represents the particle-hole channel of the nucleon-nucleon interaction. In this channel, the four nucleon operators of the interaction are coupled in the manner depicted schematically by the diagram

$$c^+ \; c^+ \; c \; c \; .$$

The matrix B_{pq} of eq. (5.9d) represents the particle-particle (or hole-hole) channel

$$c^+ \; c^+ \; c \; c \; .$$

In the case of the pairing force, the particle-hole channel is often neglected, i.e. one assumes that $A_{pq} = 0$ for all p and q. With this assumption, the conditions (5.9a,b) lead to the BCS pairing equations. A more general treatment is given below (the main results were given previously by Kumar et al. 1977).

For the pairing interaction of eq. (5.2a), the matrices A and B become diagonal and we obtain

$$A_{pq} = -G v_p^2 \delta_{pq}, \qquad (5.10a)$$

$$B_{pq} = \Delta \delta_{pq}, \qquad (5.10b)$$

where $\Delta = G \sum_p' u_p v_p$. (5.10c)

Note that we have employed the phase convention

$$u_{\bar{p}} = u_p, \quad v_{\bar{p}} = v_p, \qquad (5.11)$$

which follows from the tensorial and the time-reversal properties of the fermion operators (Kumar 1975 b). On substituting eqs. (5.10a,b) into (5.9a,b) and working in a basis where H_{av} is diagonal, i.e. eq. (5.2b) is valid, one obtains

$$2u_p v_p \left(\eta_p - \lambda - G v_p^2\right) - \Delta\left(u_p^2 - v_p^2\right) = 0, \qquad (5.12a)$$

and $\left(u_p^2 - v_p^2\right)\left(\eta_p - \lambda - G v_p^2\right) + 2\Delta u_p v_p = E_p$. (5.12b)

The Bogolyubov conditions (5.12a,b) are solved as follows. First, we eliminate the $(\eta_p - \lambda - G v_p^2)$ term by substracting $(u_p^2 - v_p^2)$ x eq. (5.12a) from $2u_p v_p$ x eq. (5.12b):

$$\Delta \left[4u_p^2 v_p^2 + \left(u_p^2 - v_p^2\right)^2\right] = 2u_p v_p E_p,$$

which becomes (on using the normalization condition 5.3c)

$$\Delta = 2u_p v_p E_p. \qquad (5.13a)$$

Similarly, by adding $2u_p v_p$ x eq. (5.12a) to $(u_p^2 - v_p^2)$ x eq. (5.12b), we obtain

$$\eta_p - \lambda - G v_p^2 = \left(u_p^2 - v_p^2\right) E_p = \left(1 - 2v_p^2\right) E_p,$$

which can be rewritten as

$$2v_p^2 = \frac{E_p - (\eta_p - \lambda)}{E_p - \tfrac{1}{2}G} = 1 - \frac{\eta_p - \lambda - \tfrac{1}{2}G}{E_p - \tfrac{1}{2}G} \quad . \tag{5.13b}$$

This, combined with eq. (5.3c), gives

$$2u_p^2 = 2 - 2v_p^2 = 1 + \frac{\eta_p - \lambda - \tfrac{1}{2}G}{E_p - \tfrac{1}{2}G} \quad . \tag{5.13c}$$

Values of λ and Δ are determined by solving simultaneously two pairing equations. One of them comes from combining eqs. (5.10c) and (5.13a) (for $\Delta \neq 0$)

$$\frac{2}{G} = \sum_p{}' \frac{1}{E_p} \quad . \tag{5.14a}$$

The second comes from the particle-number condition (5.6), applied to the zero-quasi-particle state,

$$N = \sum_p{}' 2v_p^2 \quad . \tag{5.14b}$$

Equations (5.14a,b) are identical to the BCS equations. However, the present expressions for v_p^2, E_p are different. The occupation probability, v_p^2, is defined by eq. (5.13b). This reduces to the BCS form only if $\tfrac{1}{2}G \ll |\eta_p - \lambda|$, that is only for levels far above or far below the Fermi energy λ. The expression for the quasi-particle energy, E_p, also differs from the BCS treatment, as shown below.

Rewrite eq. (5.13a) as

$$\Delta/E_p = 2u_p v_p$$
$$= \sqrt{(2u_p^2)(2v_p^2)} \quad ,$$

which on employing eqs. (5.13b,c) becomes

$$\frac{\Delta}{E_p} = \sqrt{1 - \frac{(\eta_p - \lambda - \tfrac{1}{2}G)^2}{(E_p - \tfrac{1}{2}G)^2}} \quad .$$

The last equation can be rewritten as

$$E_p = \tfrac{1}{2}G + \frac{|\eta_p - \lambda - \tfrac{1}{2}G|}{\sqrt{1 - \Delta^2/E_p^2}} \quad . \tag{5.15}$$

This is an iterative equation in E_p, which looks quite different from the BCS equation

$$(E_p)_{BCS} = \sqrt{(\eta_p - \lambda)^2 + \Delta^2} \quad . \tag{5.16}$$

However, if we substitute eq. (5.16) into the right-hand-side of eq. (5.15) and if we neglect the $\frac{1}{2}G$ terms, then eq. (5.15) becomes an identity. In practice, we find that the BCS eq. (5.16) provides a reasonable initial guess and that the iteration, implicit in eq. 5.15, converges quite rapidly after 1-5 iterations.

Thus, without introducing much extra complication or increase of computation time, the above treatment allows us to improve the pairing theory quite significantly. The present treatment is more satisfying conceptually, because the self-energy term ($-G\, v_p^2$) is included explicitly. Employing the standard BCS theory, such a self-energy term has been computed previously (Lien et al. 1975). The self-energy term, found to vary from 0.11 to 0.25 MeV for different single-particle states in $^{123-131}$Te, is comparable to level spacings.

The present treatment leads to a dramatic improvement of the pairing theory in the limit of a vanishing energy gap ($\Delta \to 0$), especially if two or more levels cross the Fermi surface at the same time. This turns out to be particularly important for the removal of a divergence in the "cranking" moments of inertia and mass parameters. Hence, we discuss in some detail the $\Delta \to 0$ limit of the pairing theory.

On substituting $\Delta = 0$ in the Bogolyubov conditions (5.12a,b), one finds that

$$2u_p v_p = 0 , \qquad (5.17a)$$

and $\quad E_p = (1 - 2v_p^2)(\eta_p - \lambda - G v_p^2) . \qquad (5.17b)$

This leads to two solutions:

either $\quad u_p = 0, \quad v_p = 1, \quad E_p = G + \lambda - \eta_p , \qquad (5.17c)$

or $\quad u_p = 1, \quad v_p = 0, \quad E_p = \eta_p - \lambda . \qquad (5.17d)$

Now, the question arises: what is the value of λ? Value of λ must satisfy the particle number condition (5.14b). But this condition, considered by itself, can be satisfied by any λ value lying between the energy of the uppermost occupied level and the lowest unoccupied level. However, we can get a precise value of λ by considering the limit $\Delta \to 0$.

Consider two levels closest to the Fermi surface. Let $\eta_>, \Omega_>$ be the energy, degeneracy of the level above λ, and let $\eta_<, \Omega_<$ denote the corresponding quantities for the level below. In the limit $\Delta \to 0$, all levels below $\eta_<$ remain totally occupied. Hence, the particle number condition implies in the present case the condition

$$\Omega_< v_<^2 + \Omega_> v_>^2 = \Omega_< . \qquad (5.18)$$

With the definitions

$$\omega^2 = \Omega_< / \Omega_> , \quad v_>^2 = \delta^2 , \qquad (5.19)$$

eq. (5.18) gives

$$v_<^2 = 1 - \delta^2/\omega^2 . \qquad (5.20)$$

Since $v_> \to 0$ as $\Delta \to 0$ and since $v_<^2$ is independent of the sign of Δ, the quantity δ is proportional to Δ. The leading order terms of the quasi-particle energies are then given by (using eq. 5.13a)

$$E_> = \frac{\Delta}{2\delta} , \quad E_< = \frac{\omega\Delta}{2\delta} . \qquad (5.21)$$

Substitution into eq. (5.13b) for the two levels, and elimination of the Δ-dependent term yields

$$\lambda = \frac{\omega\eta_> + (\eta_< - G)}{\omega + 1} , \qquad (5.22)$$

while elimination of the λ-term gives

$$E_> = \frac{\eta_> - \eta_< + G}{\omega + 1} = \eta_> - \lambda , \qquad (5.23a)$$

and $\quad E_< = \frac{\omega(\eta_> - \eta_< + G)}{\omega + 1} = \lambda + G - \eta_< . \qquad (5.23b)$

Note that eqs. (5.23a,b) are consistent with eqs. (5.17c,d) although the later equations were not employed in the above derivation.

Besides giving a precise value of λ, the above procedure also tells us what happens if the two levels happen to cross each other in the region of λ, that is if the particle-hole gap $(\eta_> - \eta_<)$ vanishes. Then, eqs. (5.23a,b) become

$$E_> = \frac{G}{\omega + 1} \xrightarrow[\omega \to 1]{} \tfrac{1}{2}G , \qquad (5.24a)$$

and $\quad E_< = \frac{\omega G}{\omega + 1} \xrightarrow[\omega \to 1]{} \tfrac{1}{2}G .. \qquad (5.24b)$

The above equations exhibit the most important consequence of the present treatment of pairing: The quasi-particle energy never vanishes, in contrast to the BCS value of eq. (5.16) which would have vanished for $\Delta = 0$ and $\eta = \lambda$ [Note that the $-G$ term of eq. (5.22) comes from the inclusion of the self-energy term in the present theory and, therefore, λ would equal η in the BCS treatment of level-crossing $(\eta_> = \eta_< = \eta)$.].

There still remains the question of choice between the two possible solutions (5.17c or d) for the u,v factors. It follows from eq. (5.22) that

$$\lambda - \eta_< > 0 \text{ if } \eta_> - \eta_< > \frac{G}{\omega} . \qquad (5.25)$$

Then, the Fermi energy lies in between the two levels, and there is no problem of interpretation: The solution (5.17c) is valid for all levels below λ, while (5.17d) applies to all levels above.

However, if the particle-hole gap is so small that $\eta_> - \eta_< \leqslant G/\omega$, then the Fermi energy coincides with $\eta_<$ or is even lower. Then, how should one choose the u,v factors? Here, we should keep in mind that the equation (5.10c) implies that if Δ equals zero, then the product $u_p v_p$ must vanish for each single-particle level (as long as we maintain the standard phase convention according to which $u_p \geqslant 0$, $v_p \geqslant 0$).

If the particle-hole gap is non-zero, we can avoid the above problem by defining an effective Fermi energy:

$$\lambda_{eff} = \lambda + \frac{G}{\omega+1} \tag{5.26}$$

Then, the solution (5.17c) is employed for all levels below λ_{eff}, while (5.17d) is employed for all levels above.

If the particle-hole gap does vanish, then we must have $v_<^2 = v_>^2$ and the particle number condition (5.18) requires

$$u = \frac{1}{\sqrt{1+\omega^2}} \quad , \quad v = \frac{\omega}{\sqrt{1+\omega^2}} \quad , \tag{5.27}$$

for the levels coinciding with λ_{eff}. This equation is in contradiction with eq. (5.10c) for $\Delta = 0$. This contradiction can be avoided by replacing the conditions (5.11) by

$$u_{\bar{p}} = u_p \quad , \quad v_{\bar{p}} = -v_p \quad , \tag{5.11'}$$

for the levels coinciding with λ_{eff} only. Since the levels are doubly degenerate, this insures the satisfaction of eq. (5.10c) with $\Delta = 0$. Although this kind of reasoning was not given, the phase convention of eq. (5.11') was already sagaciously proposed by Belyaev (1959).

B. Determination of Global Parameters

Usually, the pairing strength parameter (actually parameters - one for protons and one for neutrons) is determined by fitting the observed odd-even mass differences for a nuclear region. However, in this method, one makes two types of assumptions: (i) Nuclear shape and energy gap do not change much from nucleus to nucleus. (ii) The ground state of an odd-A nucleus is a pure one-quasi-particle state, while that of an even-even nucleus is a pure zero-quasi-particle state.

These assumptions can be dropped in an alternative method based on the extra binding energy of two particles (holes) near a doubly-closed-shell nucleus (Bohr & Mottelson 1975). This problem is a two-particle problem and, hence, the pairing interaction Hamiltonian (eq. 5.1b combined with eq. 5.2a) can be solved exactly within a given configuration space.

The trial wave function is written as

$$|\Psi\rangle = \sum_{j\Omega}{}' A_{j\Omega} |\rangle_0 , \qquad (5.28)$$

where $|\rangle_0$ is a paired state containing two nucleons in time-reversed orbits. Matrix elements of the pairing Hamiltonian in this basis are given by

$$H_{j\Omega,k\Omega'} = \begin{Bmatrix} 2\eta_j - G & \text{if } k=j, \Omega'=\Omega \\ -G & \text{otherwise} \end{Bmatrix} , \qquad (5.29)$$

where η_j is the positive energy difference between the level j and the unperturbed level (the 2-particle or 2-hole level in the absence of pair interaction). Note that since the shape is spherical for the ground state of a doubly-closed-shell nucleus, the level energy is independent of Ω. Then, the eigenvalue equation is given by

$$(2\eta_j - G - E) A_{j\Omega} + \sum_{k\Omega' \neq j\Omega}{}' (-G) A_{k\Omega'} = 0 , \qquad (5.30a)$$

which can be rewritten as

$$(2\eta_j - E) A_{j\Omega} = G \sum_{k\Omega'}{}' A_{k\Omega'} . \qquad (5.30b)$$

With the definition

$$K = G \sum_{k\Omega'}{}' A_{k\Omega'} , \qquad (5.31a)$$

eq. (5.30b) gives

$$A_{j\Omega} = K/(2\eta_j - E) . \qquad (5.31b)$$

On combining eqs. (5.31a,b), one gets the dispersion relation,

$$\frac{1}{G} = \sum_{j\Omega}{}' \frac{1}{2\eta_j - E} = \sum_j \frac{j + \tfrac{1}{2}}{2\eta_j - E} . \qquad (5.32)$$

The RPA type of equation (5.32) gives many solutions. The lowest of these is identified with the extra binding energy of two particles (holes) near a closed shell. Taking this (E) from the nuclear mass tables, eq. (5.32) is solved for the pairing strength G. The strength value obtained in this way depends on the number of levels included in the summation in eq. (5.32). Within the restriction that the number of levels above and below the Fermi surface should be equal, the level number was varied. It was found (Kumar et al. 1977) that the method converges quite well if all levels below the Fermi surface (and an equal number above) are included in the calculation.

Such a calculation was performed for 12 nuclei (closed shell ± 2 neutrons or protons) near ^{16}O, ^{40}Ca, and ^{208}Pb. It was found that the neutron and proton pairing strengths are almost equal. However, it has been shown by Greiner (1966) that a substantially larger proton pairing strength is needed in order to explain the lowering of the observed gyromagnetic ratio below the expected value Z/A. Hence, we employ the Bohr-Mottelson formulation of the Z-A dependence of the pair strengths,

$$G_{p \atop n} = G_0 \left[1 \pm G_1 \frac{N-Z}{A} \right] , \quad (5.33)$$

where G_0 is the isoscalar constant and G_1 is the isovector constant. Then, we fix the G_1 value (which determines the ratio G_p/G_n) by fitting the gyromagnetic ratio (half the magnetic moment of the first 2^+ state) of a heavy, well-deformed nucleus like ^{168}Er. And then, we use the fit to the extra binding energy, mentioned above, to determine G_0. This procedure gave (Kumar et al. 1977)

$$G_0 = 17 \text{ MeV/A} . \quad (5.34)$$

In the earlier calculations (Kumar et al. 1977, Kumar 1978), a value of $G_1 = 0.3$ was employed. However, the removal of an error in the pairing treatment of the gyromagnetic ratio led to a redetermination of G_1 and its value was changed to

$$G_1 = 0.5 \text{ MeV} . \quad (5.35)$$

The pair strength parameters of eqs. (5.34,35) have been employed for all nuclei included in the global study (Kumar 1979a) and in all subsequent calculations (Kumar 1979b, Lagrange et al. 1980, Aase et al. 1980, Vågnes et al. 1980, and to be published).

VI. Potential Energy of Deformation

A. Strutinsky Method

The shell-correction method of Strutinsky (1966) allows us to compensate for the uncertainties/errors in the single-particle levels far from the Fermi surface.

The potential energy of deformation is written as

$$V_S = V_{DM} + \delta U + \delta V_p , \qquad (6.1)$$

where V_{DM} (the droplet or liquid-drop model energy) represents the energy due to uniformly distributed single-particle levels, δU represents the shell-correction energy due to the non-uniform levels of the microscopic (Rainwater-Nilsson, in our case) Hamiltonian, and δV_p is the pair-correction energy. Our calculation of V_{DM} is discussed in sect. 6B. The calculation of the other two terms is discussed below.

Since the droplet model energy V_{DM} gives only the "smooth" part of the shape-dependence of V_S, and since δV_p also varies smoothly with nuclear shape, the important variations in the average nuclear shape are attributed to the shell-correction δU. Hence, the calculation of δU represents the most crucial step of the present program.

The shell-correction energy is defined as the difference between the energy due to the actual single-particle levels and that due to a uniform distribution of the levels:

$$\delta U = U - \bar{U} , \qquad (6.2)$$

where
$$U = 2 {\sum_p}' \eta_p \qquad (\eta_p < \lambda) , \qquad (6.3)$$

$$\bar{U} = \int_0^N \bar{\eta}(n) \, dn , \qquad (6.4)$$

λ represents the Fermi energy below (above) which all levels are occupied (unoccupied) in the absence of pairing, N is the number of nucleons (neutrons or protons), and $\bar{\eta}$ is a smooth function of particle number (n).

The smooth energy $\bar{U}$ is calculated by first replacing the n-integral by an energy-integral

$$\bar{U} = \int_{-\infty}^{\bar{\lambda}} \eta \, \bar{\rho}(\eta) \, d\eta , \qquad (6.5)$$

where $\bar{\rho}(\eta)$ is the uniform-level density (number of uniform levels per unit energy), and $\bar{\lambda}$ is the uniform Fermi energy, determined by the particle number condition

$$\bar{N} = \int_{-\infty}^{\bar{\lambda}} \bar{\rho}(\eta) \, d\eta . \qquad (6.6)$$

The uniform-level density is obtained by smoothing each single-particle level via a Gaussian centered at each level, multiplied by a polynomial in the energy distance from the level. The argument is as follows. The quantized, non-uniform level density is given by

$$\rho(\eta) = 2 \sum_p{}' \delta(\eta - \eta_p)$$

$$= 2 \sum_p{}' \frac{1}{\gamma\sqrt{\pi}} \sum_{k=0,2,\ldots}^{\infty} a_k H_k(u_p) \exp(-u_p^2) , \qquad (6.7)$$

where $u_p = (\eta - \eta_p)/\gamma$, $a_k = -a_{k-2}/(2k)$, $a_0 = 1$. $\qquad (6.8)$

and γ is a width parameter. These equations are based on a general, exact expansion of a δ-function in terms of a product of a Gaussian and a Hermite-polynomial H_k. Now, the uniform-level density is obtained by rewriting eq. (6.7) but restricting the sum to $k=0,2,\ldots,\ell$. If the number ℓ is too large (or γ is too small), one approaches the quantized level density. If the number ℓ is too small (or γ is too large), then all information of the quantized levels is lost. Experience with different sets led to the prescription (Brack et al. 1972, Nix 1972) $\ell = 6$ and $\gamma \approx \hbar\omega$. Then, the uniform-level density is given by

$$\bar{\rho}(\eta) = \frac{2}{\gamma\sqrt{\pi}} \sum_p{}' G(u_p^2) \exp(-u_p^2) , \qquad (6.9)$$

where $G(u_p^2) = \frac{1}{48}(105 - 210 u_p^2 + 84 u_p^4 - 8 u_p^6)$. $\qquad (6.10)$

With the density function given by eqs. (6.9, 6.10), the integrals of eqs. (6.5, 6.6) can be expressed in terms of the error function

$$\mathrm{erf}(x) = \frac{2}{\sqrt{\pi}} \int_0^x dz \exp(-z^2) , \qquad (6.11)$$

and are given by

$$\bar{N} = \sum_p{}' f(x_p) , \quad \left(x_p = \frac{\bar{\lambda} - \eta_p}{\gamma}\right) , \qquad (6.12)$$

$$\bar{U} = \sum_p{}' [\eta_p f(x_p) - g(x_p)] , \qquad (6.13)$$

where $f(x) = 1 + \mathrm{erf}(x) + \frac{x}{48\sqrt{\pi}} \exp(-x^2)(114 - 64 x^2 + 8 x^4)$, $\qquad (6.14)$

and $\quad g(x) = \frac{\gamma}{48\sqrt{\pi}} \exp(-x^2)(15 - 90 x^2 + 60 x^4 - 8 x^6)$. $\qquad (6.15)$

The width parameter γ is usually expressed in terms of the oscillator energy,

$$\gamma = \Gamma(\hbar\omega) . \qquad (6.16)$$

As mentioned above, the quantity Γ is expected to be of the order of unity. Its value is determined, in principle, by the convergence condition

$$\frac{\partial}{\partial \Gamma}[\delta U(\Gamma)] = 0 . \qquad (6.17)$$

However, in practice, this condition is not satisfied very well – especially in transitional nuclei where a few percent change in Γ can change a spherical nucleus into a deformed nucleus and vice versa.

Hence, a number of prescriptions have been developed over the years. In most nuclei, it suffices to replace the condition (6.17) by a weaker condition – the so-called "plateau" condition, where one searches for a local convergence in the neighborhood of $\Gamma \sim 1.0$. This plateau condition is expected to yield an accuracy of about 0.5 MeV. This is adequate for rigid-spherical nuclei (for which the expected shell-correction varies from -5 to -10 MeV) or for rigid-deformed nuclei (for which δU varies from 5 to 10 MeV), but the accuracy is not adequate for the large number of soft or transitional nuclei. Hence, we have attempted a number of prescriptions for our global study of different even-even nuclei.

In the 1977 version of the dynamic deformation theory (Kumar et al. 1977), the following prescription was employed: For $\Gamma = 1.0$, calculate $V_S(\beta,\gamma)$ and search for β_{min}, γ_{min} corresponding to the lowest minimum of V. Then, search for the plateau of $\delta U(\beta_{min}, \gamma_{min}, \Gamma) - \delta U(0, 0, \Gamma)$, i.e. for the deformation energy due to shell-correction. This prescription was employed for three well-deformed nuclei: ^{24}Mg, ^{102}Zr, and ^{168}Er. Some results were also presented for several transitional nuclei. Although the general features of spherical-deformed transition at N = 88-90, prolate-oblate transition in the Os region, and of shape as well as gap transitons in the Hg nuclei were reproduced, the calculated spectra of transitional nuclei were not satisfactory.

In the 1978 version of the theory (Kumar 1978) applied to Ge nuclei, the width parameter Γ was employed as a free parameter whose value was determined by fitting the energy of the first 2^+ state. A number of other modifications of the theory were also reported.

In the 1979I version of the theory (Kumar 1979a), a global study of even-even nuclei ranging from ^{12}C to ^{240}Pu was attempted where the Γ value was fixed via the relation

$$\Gamma = 0.3 + 0.2A^{1/3} . \qquad (6.18)$$

This relation was based on the Γ values obtained via the convergence condition (6.17) applied to two well-deformed nuclei, ^{24}Mg and ^{168}Er.

It was shown that the general features of the low-energy spectra of different types of nuclei (spherical-transitional-deformed as well as light-medium-heavy) could be reproduced without any free parameters. However, the accuracy was not quite satisfactory for transitional nuclei like ^{56}Fe and ^{154}Gd. Hence, another prescription for the determination of Γ was reported in the same paper.

In the 1979II version of the theory (Kumar 1979a), the convergence condition (6.17) was replaced by the plateau condition

$$\sigma_\Gamma(<\delta U>) = 0 \quad (\Gamma \sim 1.0, \Delta\Gamma \sim 0.1) , \qquad (6.19)$$

where $<\delta U> = \iint \delta U(\beta,\gamma) \, w(\beta,\gamma) \, d\beta \, d\gamma$, (6.20)

and σ_Γ denotes mean square deviation with respect to changes in Γ by $\Delta\Gamma$. The integral in eq. (6.20) is of the same type which appears in the solution of the collective Schrödinger equation (discussed above in ch. III). The weight factor $w(\beta,\gamma)$ is chosen in accordance with the vibrational wave function for the ground state and is given by

$$w(\beta,\gamma) = F(\beta,\gamma) \, A_{000}^2 (\beta,\gamma) ,$$

which on using eqs. 3.72, 3.73, 3.79 becomes

$$w(\beta,\gamma) = \frac{4}{b\sqrt{\pi}} \left(\frac{\beta}{b}\right)^4 \sin 3\gamma \, \exp\left(-\frac{\beta^2}{b^2}\right) . \qquad (6.21)$$

The range parameter "b" is taken from a global study of the observed B(E2) values (see eq. 3.86g).

While it is true that most nuclei are not vibrational, the above weight-function is reasonable for our global study because
 (i) the weight-function vanishes at $\beta = 0$ and peaks at $\beta = b\sqrt{2} = 1.2\sqrt{2/A}$, and
 (ii) the spectrum of a nucleus depends not only on $\delta U(\beta_{min}, \gamma_{min})$ but also on the changes in δU compared to other shapes (particularly the spherical-deformed difference and the prolate-oblate difference).

VI. B. The Droplet Model

The droplet model (Myers and Swiatecki 1969, 1974; Myers 1976) is an extension of the liquid drop model, where the nuclear energy is expanded not only in powers of $A^{-1/3}$ but also in powers of I^2, where $I=(N-Z)/A$. This model predicts not only the nuclear masses and fission barriers, but also may be applied to predictions of nuclear radii, isotope shifts, the neutron skin thickness, and to the prediction of nuclear potential-well parameters.

In recent versions of the droplet model, the nuclear mass energy is expressed as a "mass excess" energy, i.e. the energy constant referring to the ^{12}C scale is subtracted out. Then, the droplet energy (including the Wigner term and the electron binding term, but not the shell correction or the even-odd term) is given by

$$V_{DM} = 8.07169 \ N + 7.28922 \ Z$$
$$+ (-15.96 + 36.8 \ \bar{\delta}^2 - 120 \ \bar{\epsilon}^2) \ A$$
$$+ (20.69 + 179.238 \ \bar{\delta}^2) \ B_s A^{2/3}$$
$$+ 0.73219 \ Z^2 B_c/A^{1/3} - 1.6302 \times 10^{-4} x Z^2 B_r A^{1/3}$$
$$- 1.28846 \ Z^2/A - 0.44377 Z - 4.9274 \times 10^{-4} x Z^2 B_w$$
$$+ 30 \ |N-Z|/A - 1.433 \times 10^{-5} x Z^{2.39}, \qquad (6.23)$$

where all the numbers are in MeV,

$$\bar{\delta} = \frac{(N-Z)/A + 0.0080756 \ Z B_v/A^{2/3}}{1 + 4.8706 \ B_s/A^{1/3}}, \qquad (6.24)$$

$$\bar{\epsilon} = -0.17242 \ B_s/A^{1/3} + 5.88235 \bar{\delta}^2$$
$$+0.0030508 \ Z^2 B_c/A^{4/3}, \qquad (6.25)$$

and B_s, B_c, B_r, B_v, B_w are the surface, coulomb, and three additional functions of nuclear shape defined by Myers and Swiatecki. (The sixth function, B_k, is not needed since the corresponding mass formula coefficient is zero.)

B. Remaud (1978) has developed an exact method of calculating the six functions B_s, B_c,... which gives the same results as the axially-symmetric-shape-dependent expansion of Hasse (1971) at small β but which is more accurate at large β (we go up to $\beta=0.8$) and which is more general (we need these functions not only for $\gamma=0°$ but for the whole range of $\gamma=0°-60°$). This exact method has been employed in the DDT calculations performed since 1978.

The values of $V_{DM}(\beta,\gamma)$ calculated via eqs. (6.23-25) are employed in the DDT calculations of the potential energy via eq. (6.1).

VI. C. The Projection Correction

The nuclear ground state energy (or the binding energy, or the excess mass) is often determined by minimizing the potential energy function [for instance, $V_s(\beta,\gamma)$ of eq. (6.1)] with respect to the shape variables (β,γ in the DDT). However, this procedure is incorrect. It neglects the energy of zero-point-motion (which would be $\frac{5}{2}\hbar\omega$ for a five-dimensional-harmonic motion).

On the other hand, simply adding the zero-point-energy to V_{min} leads to an inconsistency with the general rule of quantum-mechanics that an improvement of the wave function lowers the energy. One way of removing this inconsistency would be to employ the projection method where the projected ground state energy is lower than the Hartree-Fock-Bogolyubov minimum. However, the projection method is too impractical for the general problem of nine collective variables.

A practical way of removing the above-mentioned inconsistency has recently been found (Kumar, 1979 a,b). We would like to restate the argument as follows. The energy V_s of eq. (6.1) is an approximation to the Hartree-Fock-Bogolyubov energy

$$V_s = \langle 0|H|0\rangle, \tag{6.26}$$

where $|0\rangle$ is the lowest HFB state and H is the microscopic Hamiltonian. Now, the HFB wavefunction represents an overlap of collective states with different angular momenta and with different vibrational quanta. In other words, H is not just the intrinsic Hamiltonian H_{int} (which gives the collective potential energy V) but it includes a collective kinetic energy $\hat{T}_{coll}$:

$$H = H_{int} + \hat{T}_{coll}, \tag{6.27}$$

$$H_{int} = H - \hat{T}_{coll}, \tag{6.28}$$

$$\begin{aligned} V &= \langle 0|H_{int}|0\rangle \\ &= V_s - \langle 0|\hat{T}_{coll}|0\rangle. \end{aligned} \tag{6.29}$$

According to eq. (2.18), the collective kinetic energy operator is given by

$$\hat{T}_{coll} = -\frac{\hbar^2}{2} D^{-\frac{1}{2}} \sum_{\mu\nu} \frac{\partial}{\partial \alpha_\mu} D^{\frac{1}{2}} B_{\mu\nu}^{-1} \frac{\partial}{\partial \alpha_\nu}, \tag{6.30}$$

where D is the determinant of the mass-parameter-matrix $B_{\mu\nu}$. On substituting eq. (6.30) into (6.29), we get

$$\begin{aligned} V &= V_s + \frac{\hbar^2}{2} D^{-\frac{1}{2}} \sum_{\mu\nu} \langle 0|\frac{\partial}{\partial \alpha_\mu} D^{\frac{1}{2}} B_{\mu\nu}^{-1} \frac{\partial}{\partial \alpha_\nu}|0\rangle \\ &= V_s + \frac{\hbar^2}{2} D^{-\frac{1}{2}} \sum_{\mu\nu} [\frac{\partial}{\partial \alpha_\mu}(D^{\frac{1}{2}} B_{\mu\nu}^{-1}) \langle 0|\frac{\partial}{\partial \alpha_\nu}|0\rangle \\ &\quad + D^{\frac{1}{2}} B_{\mu\nu}^{-1} \langle 0|\frac{\partial}{\partial \alpha_\mu}\frac{\partial}{\partial \alpha_\nu}|0\rangle] \end{aligned}$$

$$V = V_s + \frac{\hbar^2}{2} \sum_{\mu\nu} B_{\mu\nu}^{-1} \langle 0 | \frac{\partial}{\partial \alpha_\mu} \frac{\partial}{\partial \alpha_\nu} | 0 \rangle, \quad (6.31)$$

where we have employed the result $\langle 0 | \frac{\partial}{\partial \alpha_\nu} | 0 \rangle = 0$. Rewrite eq. (6.31) as

$$V = V_s + \delta E_{proj}, \quad (6.32)$$

where $\delta E_{proj} = \frac{\hbar^2}{2} \sum_{\mu\nu} B_{\mu\nu}^{-1} \langle 0 | \frac{\partial}{\partial \alpha_\mu} \frac{\partial}{\partial \alpha_\nu} | 0 \rangle$

$$= -\frac{\hbar^2}{2} \sum_{\mu\nu} B_{\mu\nu}^{-1} \sum_{n \neq 0} \frac{\langle 0 | \frac{\partial H}{\partial \alpha_\mu} | n \rangle \langle n | \frac{\partial}{\partial \alpha_\nu} | 0 \rangle}{(E_n - E_o)^2}. \quad (6.33)$$

The correction term, δE_{proj}, of eqs. (6.32, 6.33) is analogous to the zero-point-energy-correction term obtained by Goeke and Reinhard (1978) in what they call a "Consistent Theory" of collective motion. However, we prefer to call it a (nine-dimensional) projection-correction term partly because it reduces in the case of one-dimensional rotation to the "projection" term given by $-\langle J^2 \rangle/(2I)$, and partly because we wish to avoid any confusion with the "conventional" zero-point-energy term coming from the action of $\hat{T}_{coll}$ on Ψ_{coll}--which equals $\frac{5}{2}\hbar\omega$ for five-dimensional-harmonic motion.

The projection-correction term, δE_{proj}, has a rather beneficial side effect. It removes the uncertainty in the quantization of a general, kinetic energy of the type of eq. (2.16)--whose mass depends on the coordinates. As pointed out by Tanabe and Sugawara - Tanabe (1976), this kind of quantization is uncertain by a constant--not an absolute constant (that would not affect the calculated spectrum of collective states), but a constant in the sense that it does not act on Ψ_{coll}. Such a constant can be shape-dependent. Hence, it can modify V as well as Ψ_{coll}. However, in the present method, such a "constant" would have no effect on the calculated spectrum, since it would first be substracted from our V of eq. (6.32), but it would then be added to the $\hat{T}_{coll}$ of our collective Hamiltonian of eq. (6.30).

It turns out that the inclusion of δE_{proj} leads not only to the aesthetically pleasing result that the final ground state energy is below the potential minimum (without any effects of $\hat{T}_{coll}$). It is absolutely essential if we do not wish to destroy the beautiful agreement of the droplet model with the observed nuclear masses--especially in light nuclei, where the magnitude of both δE_{proj} and E_{ZPM} (energy of zero-point-motion) can be as much as several tens of MeV. (But the two have opposite signs). In fact, this method has led in some cases to even a better agreement with experiment than the droplet model (Kumar 1979b).

VII. Moments of Inertia and Mass Parameters

A. Cranking Formulae

The cranking method was employed in ch. II to derive the general expression (2.14) for the mass-parameter matrix $B_{\mu\nu}$. It was stated at the end of ch. II that nine-dimensional collective motion is included in the DDT. The corresponding collective coordinates are: β_a (a=0,2'; eq. 3.5), w_k (k=1, 2, 3; eq. 3.34), Δ_τ (τ=p,n), and λ_τ. The last four are not completely independent of the others, since they must satisfy the Bogolyubov conditions (5.7 a,b) at all times. Their time-dependences are given by

$$\dot{\Delta}_\tau = \sum_a \dot{\beta}_a \, (\partial \Delta_\tau / \partial \beta_a) \, , \tag{7.1a}$$

$$\dot{\lambda}_\tau = \sum_a \dot{\beta}_a \, (\partial \lambda_\tau / \partial \beta_a) \, . \tag{7.1b}$$

The improved pairing theory given in ch. V (and, also the improved definition of nuclear axis lengths in terms of nuclear shapes, see eqs. 4.4, 4.5) leads to several important corrections to the usual expressions for the moments of inertia and the mass parameters. Hence, some of the details of the derivation are given below.

Our time-dependent Hamiltonian (see Eq. 2.10) takes the form

$$H'' = H' - \frac{\hbar}{i} \sum_a \dot{\beta}_a \left[\frac{\partial}{\partial \beta_a} + \sum_\tau \left(\frac{\partial \Delta_\tau}{\partial \beta_a} \frac{\partial}{\partial \Delta_\tau} + \frac{\partial \lambda_\tau}{\partial \beta_a} \frac{\partial}{\partial \lambda_\tau} \right) \right] - \hbar \sum_k w_k J_k, \tag{7.2}$$

where H' is the constrained microscopic Hamiltonian defined by Eqs. (5.1-5.6) and J_k is the microscopic-angular-momentum operator. In order to calculate the mass-parameter-matrix, we follow the procedure outlined in ch. II and get the general expression of Eq. (2.14) with the following substitutions:

$$\alpha_\mu \equiv \beta_a, \quad \frac{\partial}{\partial \alpha_\mu} \equiv \frac{\partial}{\partial \beta_a} + \sum_\tau \left(\frac{\partial \Delta_\tau}{\partial \beta_a} \frac{\partial}{\partial \Delta_\tau} + \frac{\partial \lambda_\tau}{\partial \beta_a} \frac{\partial}{\partial \lambda_\tau} \right) = \frac{\partial}{\partial d_a} \, , \tag{7.3a}$$

$$\alpha_\mu \equiv \theta_k, \quad \frac{\partial}{\partial \alpha_\mu} \equiv J_k, \tag{7.3b}$$

where θ_k are some angles whose time-dependences are related to those of the Euler's angles via Eqs. (3.34).

The operator J_k changes sign under time-reversal, but the operator $\partial/\partial d_a$ is invariant under time-reversal. Hence, the 5x5 mass-parameter-matrix of Eq. (2.14) splits into two unconnected parts. Of these two, the 3x3 rotational part has only diagonal matrix elements which are proportional to the three moments of inertia (k = 1, 2, 3)

$$\mathcal{J}_k = 2\hbar^2 \sum_{m,n} \frac{|<m|J_k|n>|^2}{E_m - E_n} \, . \tag{7.4}$$

The 2x2 vibrational mass-parameter-matrix is given by (a=0,2')

$$B_{ab} = 2\hbar^2 \sum_{m,n} \frac{|<n|\frac{\partial}{\partial d_a}|m>| \; |<m|\frac{\partial}{\partial d_b}|n>|}{E_m - E_n} . \qquad (7.5)$$

The matrix elements of type $<m|\frac{\partial}{\partial d}|n>$ are determined as follows. Make a Taylor series expansion of the Hamiltonian (whose solutions are $|n>, |m>, \ldots$)

$$H'(d+\delta d) = H'(d) + \frac{\partial H'}{\partial d} \delta d + \ldots \qquad (7.6)$$

Perturbation theory gives the corresponding expansion of the wave function as

$$|n(d+\delta d)> = |n(d)> - \delta d \sum_{m \neq n} \frac{|m><m|\frac{\partial H'}{\partial d}|n>}{E_m - E_n} + \ldots \qquad (7.7)$$

Then, the partial derivative of the wave function is

$$\frac{\partial |n>}{\partial d} = \lim_{\delta d \to 0} \left[\frac{|n(d+\delta d)> - |n(d)>}{\delta d}\right]$$

$$= -\sum_{m \neq n} \frac{|m><m|\frac{\partial H'}{\partial d}|n>}{E_m - E_n} , \qquad (7.8)$$

and the matrix element is given by

$$<m|\frac{\partial}{\partial d}|n> = - \frac{<m|\frac{\partial H'}{\partial d}|n>}{E_m - E_n} . \qquad (7.9)$$

Note that this is an exact result, although first order perturbation theory was employed. Thus, Eq. (7.5) can be rewritten as

$$B_{ab} = 2\hbar^2 \sum_{m,n} \frac{|<n|\frac{\partial H'}{\partial d_a}|m>| \; |<m|\frac{\partial H'}{\partial d_b}|n>|}{(E_m - E_n)^3} . \qquad (7.10)$$

The expressions (7.4) and (7.10) are quite general. They could be employed for odd-A nuclei (n=1QP, 3QP,...), odd-odd nuclei (n=2QP, 4 QP,...), or even-even nuclei (n=0 QP, 2QP,...). Now we consider the case of the low-energy states of even-even nuclei. Then, the only states which appear in the sums in Eqs. (7.4, 7.10) are

$$|n> \equiv |0QP> \equiv |0> , \qquad (7.11a)$$

$$|m> \equiv |2QP> \equiv |pq> , \qquad (7.11b)$$

where p,q denote single-quasi-particle states.

Matrix elements of the operators J_k are given by (see, e.g., Kumar 1975b)

$$s_q \langle p\bar{q}|J_k|0\rangle = (u_p v_q - v_p u_q) \langle p|J_k|q\rangle , \qquad (7.12)$$

and the energy denominator by

$$E_m - E_n = E_{pq} - E_0 = E_p + E_q - 2Gu_p v_p u_q v_q , \qquad (7.13)$$

where the term proportional to G appears due to our inclusion of the particle-hole channel of the pairing interaction (see ch. V). Note that the matrix element of Eq. (7.12) vanishes for $q=p$ or for $q=\bar{p}$ (since $u_p = u_{\bar{p}}$, $v_p = v_{\bar{p}}$), and that Eq. (7.13) applies only to the case $q \neq p$ or $\bar{p}$. Thus, the final expression for the moments of inertia is

$$\mathcal{J}_k = 2\hbar^2 \sum_{\substack{q \neq p, \bar{p} \\ p \neq q, \bar{q}}} \frac{(u_p v_q - v_p u_q)^2 |\langle p|J_k|q\rangle|^2}{E_p + E_q - 2Gu_p v_p u_q v_q} . \qquad (7.14)$$

B. Effects of Pair Fluctuations

In order to calculate the mass-parameter-matrix B_{ab}, we need to consider the operator $\partial H'/\partial d_a$, where $\partial/\partial d_a$ is defined in Eq. (7.3a). Here H' indicates the "average" of the many-body Hamiltonian (The total many-body Hamiltonian has no explicit dependence on the collective degrees of freedom.), where the "average" may be obtained in two different ways: Averaging over the particle-hole channels leads to terms proportional to $c^+_p c_q$, whose matrix elements give $H_{pq} - \lambda \delta_{pq} + A_{pq}$ of Eqs. 5.9, 5.10. Averaging over the particle-particle channels leads to terms proportional to $c^+_p c^+_q + c_p c_q$, whose matrix elements give B_{pq} of Eqs. 5.9, 5.10. Thus, we get

$$\frac{\partial H'}{\partial d_a} = \sum_{pq} \langle p|\frac{\partial H_{av}}{\partial \beta_a}|q\rangle c^+_p c_q - \sum_\tau (\frac{\partial \Delta_\tau}{\partial \beta_a} \hat{P}_\tau + \frac{\partial \lambda_\tau}{\partial \beta_a} \hat{N}_\tau) , \qquad (7.15)$$

where $\hat{P}_\tau$ is the "pair" operator,

$$\hat{P}_\tau = \sum_p' s_p (c^+_p c^+_{\bar{p}} + c_p c_{\bar{p}}) , \qquad (7.16)$$

and $\hat{N}_\tau$ is the "number" operator of Eq. (5.5).

Using the quasi-particle transformation of Eqs. (5.3) and the anticommutation rules, the needed matrix elements are given by

$$s_q \langle p\bar{q}|\frac{\partial H_{av}}{\partial \beta_a}|0\rangle = (u_p v_q + v_p u_q) \langle p|\frac{\partial H_{av}}{\partial \beta_a}|q\rangle , \qquad (7.17)$$

$$s_q <p\bar{q}| \hat{P}_\tau |0> = (u_p^2 - v_p^2) \delta_{pq}, \qquad (7.18)$$

and

$$s_q <p\bar{q}| \hat{N}_\tau |0> = 2u_p v_p \delta_{pq}. \qquad (7.19)$$

Eqs. (7.15-7.19) are combined to obtain

$$s_q <p\bar{q}|\frac{\partial H'}{\partial d_a}|0> = (u_p v_q + v_p u_q)<p|\frac{\partial H_{av}}{\partial \beta_a}|q>$$

$$- \delta_{pq}[\frac{\partial \Delta_\tau}{\partial \beta_a}(u_p^2 - v_p^2) + \frac{\partial \lambda_\tau}{\partial \beta_a}(2u_p v_p)]. \quad (7.20)$$

The derivative, $\partial H_{av}/\partial \beta_a$, is written in terms of the monopole and quadrupole operators (by using Eqs. 4.12-4.15),

$$\frac{\partial H_{av}}{\partial \beta_a} = \hbar w_{0\tau}(\frac{\partial \beta_{00}}{\partial \beta_a}\rho^2 - \frac{\partial \beta_0^P}{\partial \beta_a}Q'_{20} - \frac{\partial \beta_2^P}{\partial \beta_a}Q'_{22'}), \quad (7.21)$$

where we have denoted the proton-neutron-dependence of the scaling factor $\hbar w_0$ explicitly. For the sake of comparison, we note that the corresponding expression for the Pairing-Plus-Quadrupole model (Kumar 1963) is

$$\frac{\partial H_{av}}{\partial \beta_a} = -\hbar w_{0\tau} Q'_{2a}. \qquad (7.22)$$

The more general form of Eq. (7.21) arises because of our use of the more general (Hill-Wheeler) relations between nuclear axis-lengths and deformations.

The deformation-dependent derivatives of the energy gaps and the Fermi-energies, needed for Eq. 7.20, are determined by differentiating both sides of Eqs. (514 a,b). For the sake of simplification, we drop the subscript τ once again, and write (Recall that G, N are constants)

$$0 = \sum_p' E_{pa}/E_p^2, \qquad (7.23)$$

$$0 = \sum_p'[(n_p - \lambda - \tfrac{1}{2}G)E_{pa} - (E_p - \tfrac{1}{2}G)(n_{pa} - \lambda_a)]/(E_p - \tfrac{1}{2}G)^2, (7.24)$$

where $E_{pa} = \partial E_p/\partial \beta_a$, $\eta_{pa} = \partial \eta_p/\partial \beta_a$, $\lambda_a = \partial \lambda/\partial \beta_a$,

and we have employed Eq. (5.13b). The derivative E_{pa} is related to Δ_a and λ_a by employing Eq. (5.15). Then, the equations are solved for Δ_a and λ_a. After some algebra, we obtain

$$\frac{\partial \Delta}{\partial \beta_a} = \Delta_a = F(\frac{\partial B}{\partial \lambda}D_a - \frac{\partial A}{\partial \lambda}C_a), \qquad (7.25)$$

$$\frac{\partial \lambda}{\partial \beta} = \lambda_a = F(\frac{\partial A}{\partial \Delta}C_a - \frac{\partial B}{\partial \Delta}D_a), \qquad (7.26)$$

where $F = (\frac{\partial A}{\partial \Delta} \frac{\partial B}{\partial \lambda} - \frac{\partial B}{\partial \Delta} \frac{\partial A}{\partial \lambda})^{-1}$, (7.27)

$$\frac{\partial A}{\partial \Delta} = -\Delta \sum_p' 1/(R_p^2 S_p),$$ (7.28)

$$\frac{\partial A}{\partial \lambda} = \sum_p' (\eta_p - \lambda - \tfrac{1}{2}G)/S_p,$$ (7.29)

$$\frac{\partial B}{\partial \lambda} = \Delta^2 \sum_p' 1/(R_p S_p),$$ (7.30)

$$\frac{\partial B}{\partial \Delta} = \Delta \frac{\partial A}{\partial \lambda},$$ (7.31)

$$R_p = E_p/(E_p - \tfrac{1}{2}G),$$ (7.32)

$$S_p = R_p [(E_p - \tfrac{1}{2}G)^3 + \tfrac{1}{2}G(\eta_p - \lambda - \tfrac{1}{2}G)^2],$$ (7.33)

$$C_a = \Delta^2 \sum_p' \frac{<p|\frac{\partial H_{av}}{\partial \beta_a}|p>}{S_p R_p},$$ (7.34)

and $D_a = \sum_p' \frac{(\eta_p - \lambda - \tfrac{1}{2}G)<p|\frac{\partial H_{av}}{\partial \beta_a}|p>}{S_p},$ (7.35)

where we have employed the method indicated by Eqs. (7.6-7.8) to replace $\partial \eta_p/\partial \beta_a$ by $<p|\frac{\partial H_{av}}{\partial \beta_a}|p>$. Note that the variation of the single-particle energies with deformation is responsible for the deformation-dependence of Δ and λ.

Eqs. (7.20, 7.21, 7.25-35) are combined with Eq. (7.10) to calcualte the mass parameters B_{00}, $B_{02'}$, $B_{2'2'}$.

VIII. Calculations of Electromagnetic Moments and Transition Probabilities

The electromagnetic moments and transition probabilities provide some of the most sensitive tests of nuclear models. Of particular importance are those moments whose signs can be measured and calculated: electric quadrupole moments, E2/M1 mixing ratios, and E0/E2 mixing ratios. This list does not include the magnetic moments, since they are positive for the collective states of even-even nuclei.

Some of the most remarkable successes of the present approach (microscopic calculation of the collective Hamiltonian) have concerned the predictions of electromagnetic moments. For instance, the change of sign of the quadrupole moment of the first 2+ state in the Os-Pt region was predicted (Kumar and Baranger 1966) and later confirmed experimentally (Glenn and Saladin 1968, Pryor and Saladin 1970). Similarly, the signs of the E2/M1 mixing ratios of W-Os-Pt nuclei were predicted (Kumar 1969). Most of these were confirmed later (Hamilton and Davies 1968, Hamilton 1969, Krane and Steffen 1971), but some discrepancies in the W isotopes were also noted (Milner et al. 1971). Also, a substantial effective charge was employed in the calculation of the moments.

No effective charge parameters (or effective gyromagnetic ratios) are employed in the present theory, where a more complete configuration space is employed. Hence, comparison with the experimental moments and transition probabilities provides a more significant test of the the theory.

The method of calculation is the same as that employed previously and it has been described in some detail (Kumar and Baranger 1967, Kumar 1975a). Only the main points are mentioned below.

The electromagnetic moments of interest are related to the reduced matrix elements as follows:

$$B(T\lambda; \alpha'I' \to \alpha I) = (2I' + 1)^{-1} M^2_{\alpha'I', \alpha I}(T\lambda), \quad (8.1)$$

$$Q^S_{\alpha I} = -[\frac{16\pi I(2I-1)}{5(I+1)(2I+1)(2I+3)}]^{\frac{1}{2}} M_{\alpha I, \alpha I}(E2), \quad (8.2)$$

$$\mu^S_{\alpha I} = [\frac{4\pi I}{3(I+1)(2I+1)}]^{\frac{1}{2}} M_{\alpha I, \alpha I}(M1), \quad (8.3)$$

$$\delta(E2/M1; i \to f) = \frac{-0.835 \ (E_\gamma \text{ in MeV})[M_{if}(E2) \text{ in e.b}]}{[M_{if}(M1) \text{ in n.m.}]}, \quad (8.4)$$

where $T\lambda \equiv E\lambda$ or $M\lambda$, I is nuclear angular momentum, α is used to distinguish between different nuclear states with the same I, $B(T\lambda)$ denotes reduced transition probability, Q^S denotes spectroscopic quadrupole moment, μ^S denotes spectroscopic magnetic moment, and $M(T\lambda)$ is the reduced matrix element for the $T\lambda$ transition.

The reduced matrix element is related to the transition operator $(T\lambda)$ (Note that we consider here only the even parity operators: E2, M1,...)

$$M_{\alpha'I', \alpha I}(T\lambda) = (-1)^{\lambda-1} <\alpha I||\mathcal{M}(T\lambda)||\alpha'I'>, \qquad (8.5)$$

and to the "full" matrix element,

$$<\alpha IM|\mathcal{M}(T\lambda,\mu)|\alpha'I'M'>$$

$$= (-1)^{I'-\lambda+M} \begin{pmatrix} I' & \lambda & I \\ M' & \mu & -M \end{pmatrix} <\alpha I||\mathcal{M}(T\lambda)||\alpha'I'>, \qquad (8.6)$$

where a 3-j symbol has been employed.

Equations (8.1-8.6) give the geometrical factors needed to calculate the observable quantities like B(E2) values, Q^s,... The structure information is contained in the reduced matrix elements.

The calculation of the reduced matrix elements is performed in the manner discussed in ch. III in connection with Bohr's collective Hamiltonian and its two extreme limits (rotational and vibrational models). But there are several important differences:

1. The total wave function is not just the rotation-vibration wave function of Eqs. (3.41-3.42). Instead, it is a product of the rotation-vibration wave function of Eq. (3.41) and the zero-quasi-particle (BCS) wave function of Eq. (5.8a).

2. The electromagnetic operator (with respect to shape variables) is not just the conventional operator of the collective model (for instance, the E2 operator of Eqs. 3.83, which is based on the assumptions of uniform charge distribution and of smallness of deformation). Instead, the microscopic definitions and calculations are employed:

$$\mathcal{M}(E2, \mu) = \sum_{i=1}^{Z} r_i^2 Y_{2\mu}(\theta_i, \phi_i), \qquad (8.7)$$

$$\mathcal{M}(M1, \mu) = \sum_{i=1}^{Z} l_{i\mu} + \sum_{\tau=n,p} g_{s\tau} \sum_{i=1}^{A} s_{i\mu}, \qquad (8.8)$$

$$\mathcal{M}(E0) = \sum_{i=1}^{Z} (r_i^2/R^2), \qquad (8.9)$$

where (r_i, θ_i, ϕ_i) give the coordinates of the ith nucleon, $l_{i\mu}$ ($s_{i\mu}$) denotes the μ-component of the orbital (spin) angular momentum of the same nucleon, and R is the nuclear radius. The "bare" values of the gyromagnetic ratios, g_{sp} and g_{sn}, are employed, and no effective charges are employed.

3. Expectation values, with respect to the microscopic parts of the wave functions, of the electromagnetic operators are calculated for each point of the β-γ mesh (see Fig. 8). Thus, the complete β-γ-dependence is taken into account and no expansions are made around the potential minimum.

4. The various β-γ-dependent moments are averaged over all relevant shapes by performing double integrations of the type of Eq. (3.53). The K-mixing implicit in the calculated wave functions (see Eq. 3.42) is employed to include $\Delta K=0,2...$ transitions.

IX. Discussions of Results (Dynamic Deformation Model)

A. Parameters of the Model

As indicated above, all the parameters of the Dynamic Deformation Model (DDM) are fixed by a global study of the general features of nuclear radii (see Eq. 4.8 for the two $\hbar\omega$ relations), spherical-single-particle levels (Kumar et al. 1977, Table I), and binding energies (see Eqs. 5.33-35 for the two pairing force strengths, and Eqs. 6.23-25 for the droplet model parameters). The Z-A-dependences of the parameters, given by these relations, are employed and no parameters are varied in order to obtain the best possible fit to the experimental spectra. No effective charges or effective gyromagnetic ratios are employed.

B. A Global Study of the Low-Energy, Low-Spin Properties of Even-Even Nuclei (A=12-240)

The methods discussed above were combined in the DDM. Deformation (β,γ) dependent wave functions were calculated for different types of even-even nuclei (spherical, transitional, deformed; light, medium, heavy) without any fitting parameters. These wave-functions were then employed to calculate the energies, B(E2) values, quadrupole and magnetic moments of selected nuclei with A=12-240 (Kumar 1979a). Some of these results are reproduced in Figures 12-18.

These results showed for the first time that the gross features and even many of the fine features of the low-energy spectra of even-even nuclei could be understood on the basis of a single model and without varying any parameter from nucleus to nucleus.

The remaining descrepancies require further improvements of the model. We will postpone this discussion until Ch. X, and discuss now the detailed spectra of a few of these nuclei.

C. Shape Co-existence

It is interesting to classify different nuclei according to their shapes. However, some caution must be exercised in assessing such a classification.

It is dangerous to determine a nuclear shape solely on the basis of the potential minimum calculated within a certain model. Aside from the uncertainties of the model, one has to consider (a) whether such a minimum is an absolute minimum (If it is a maximum with respect to another degree of freedom, then it is a saddle point and not a true minimum. This is the case for most of the oblate "minima" and for the fisssion barriers.)?, and (b) whether the minimum is stable against dynamic effects (In many transitional nuclei, the spherical minimum or the shallow-deformed minimum is completely washed out by the energy of zero-point-motion.)?

Nuclear shape can be quite different in different states of the same nucleus. This phenomenon is called "shape co-existence". Three examples of this phenomenon are discussed as further tests of the DDM (Kumar 1979b).

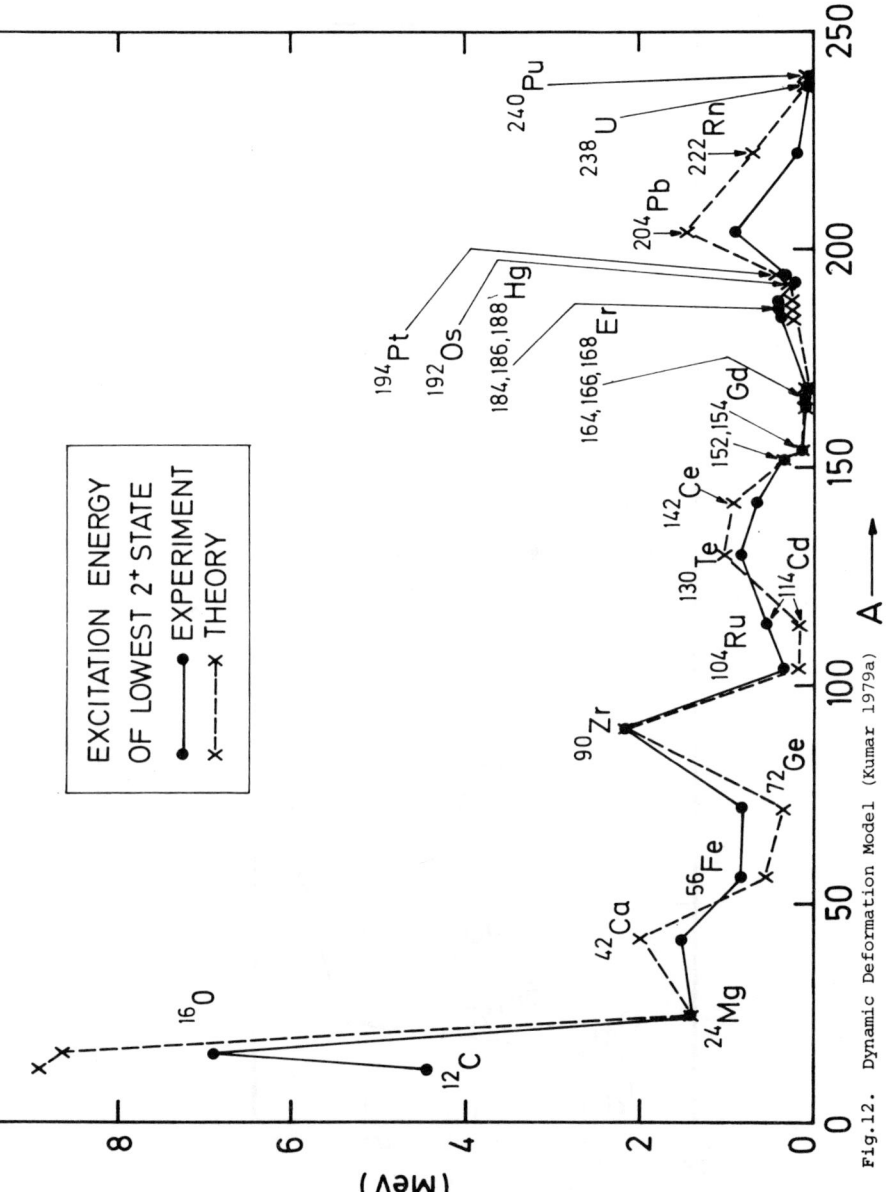

Fig.12. Dynamic Deformation Model (Kumar 1979a)

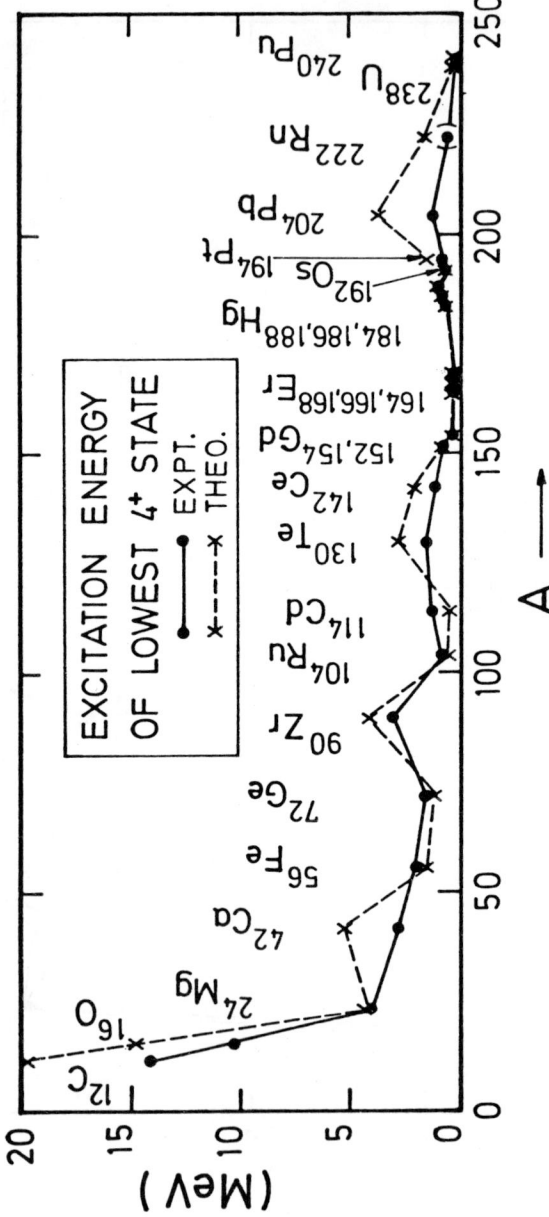

Fig.13. Dynamic Deformation Model (Kumar 1979a).

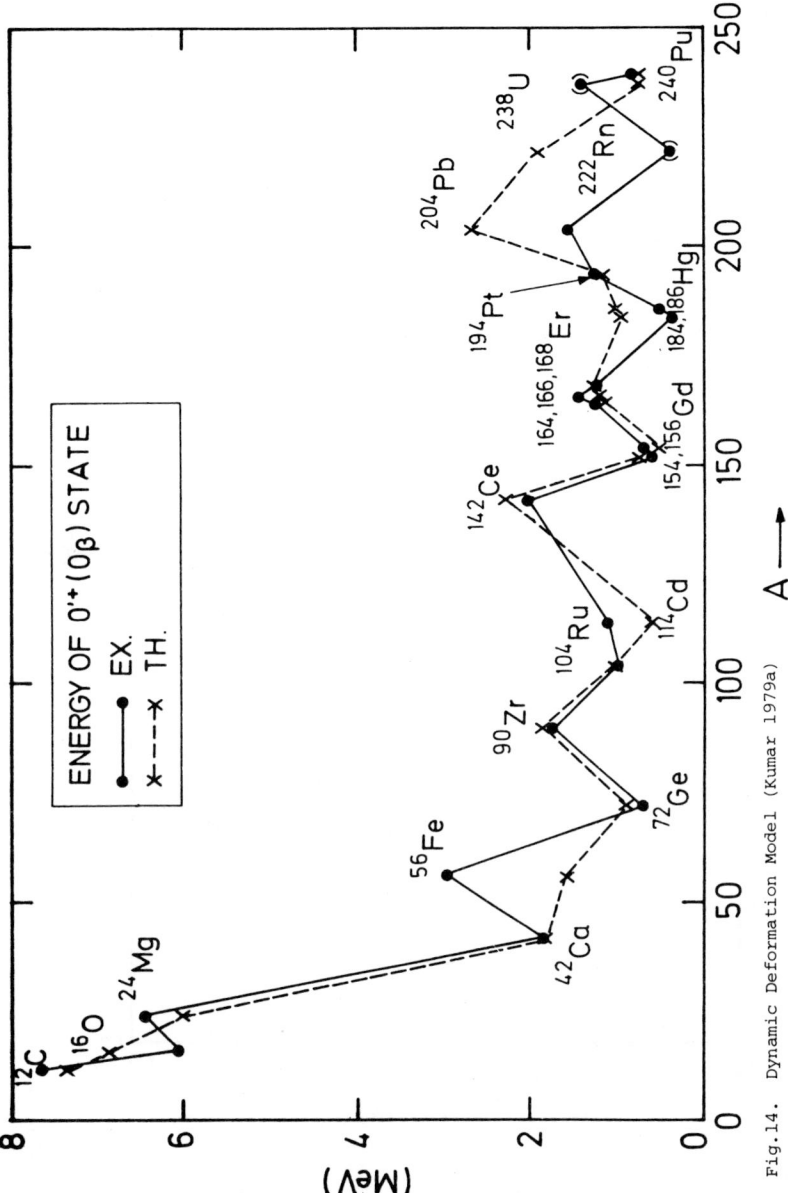

Fig.14. Dynamic Deformation Model (Kumar 1979a)

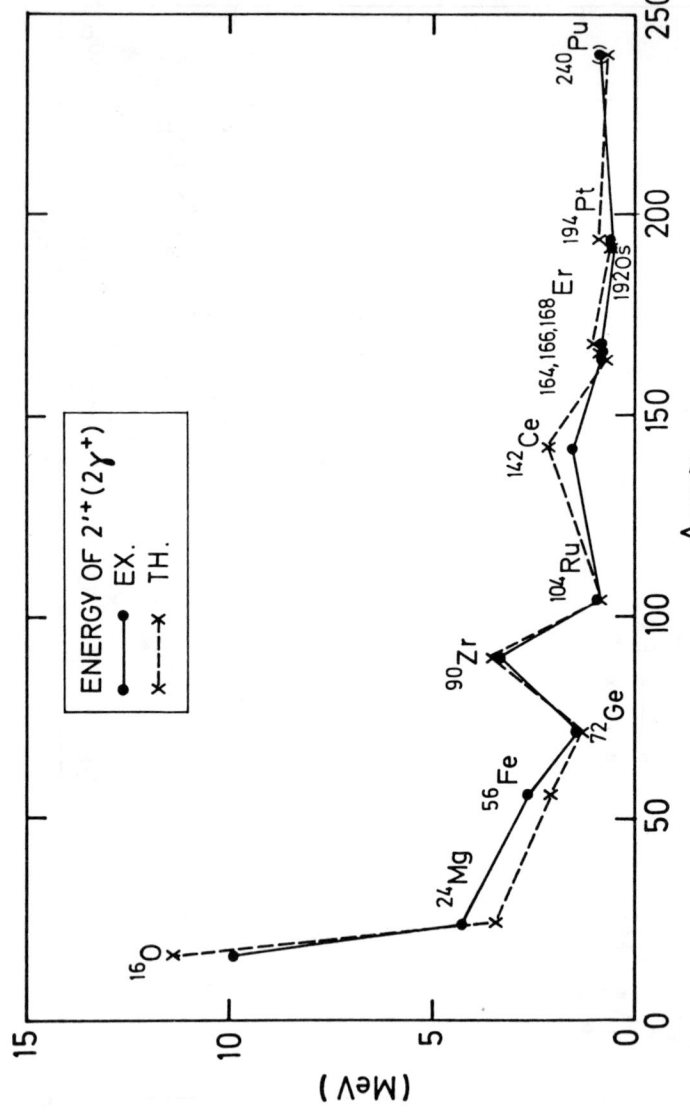

Fig.15. Dynamic Deformation Model (Kumar 1979a).

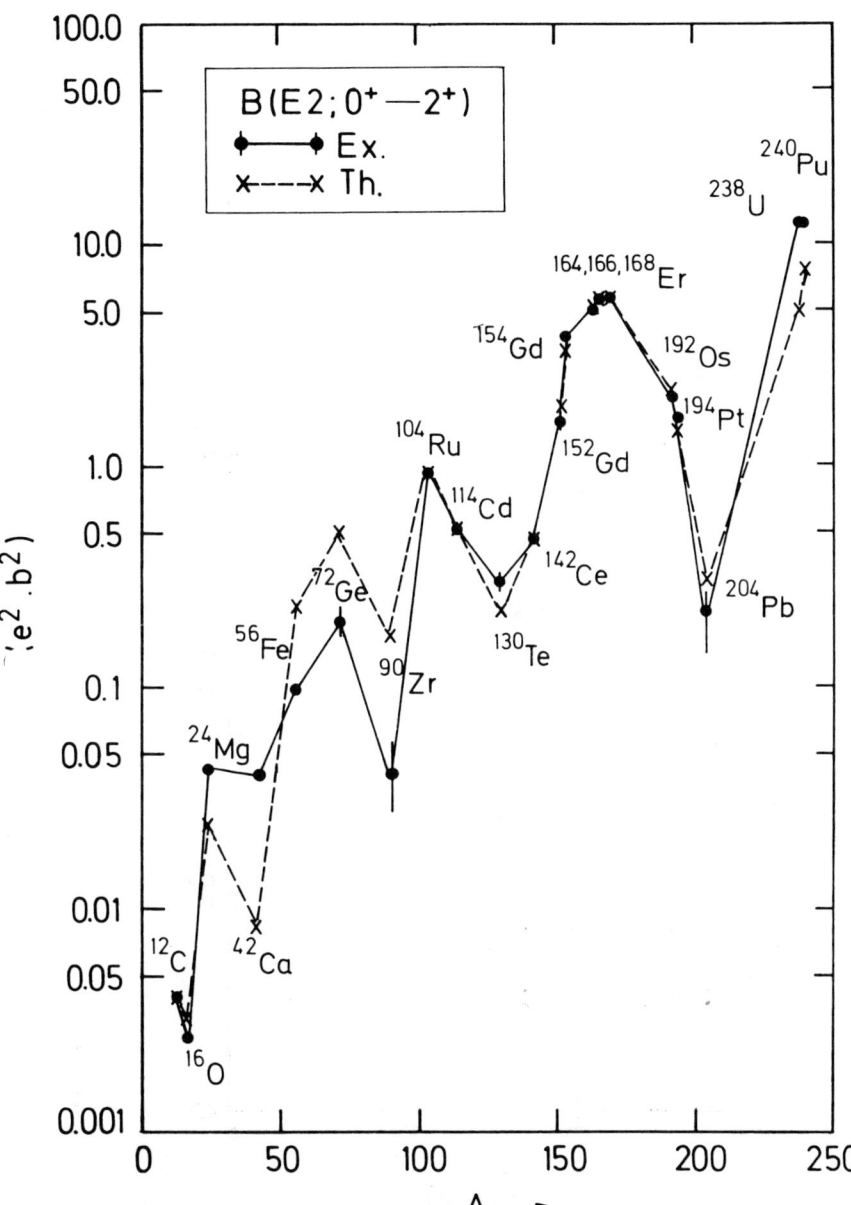

Fig.16. Dynamic Deformation Model (Kumar 1979a).

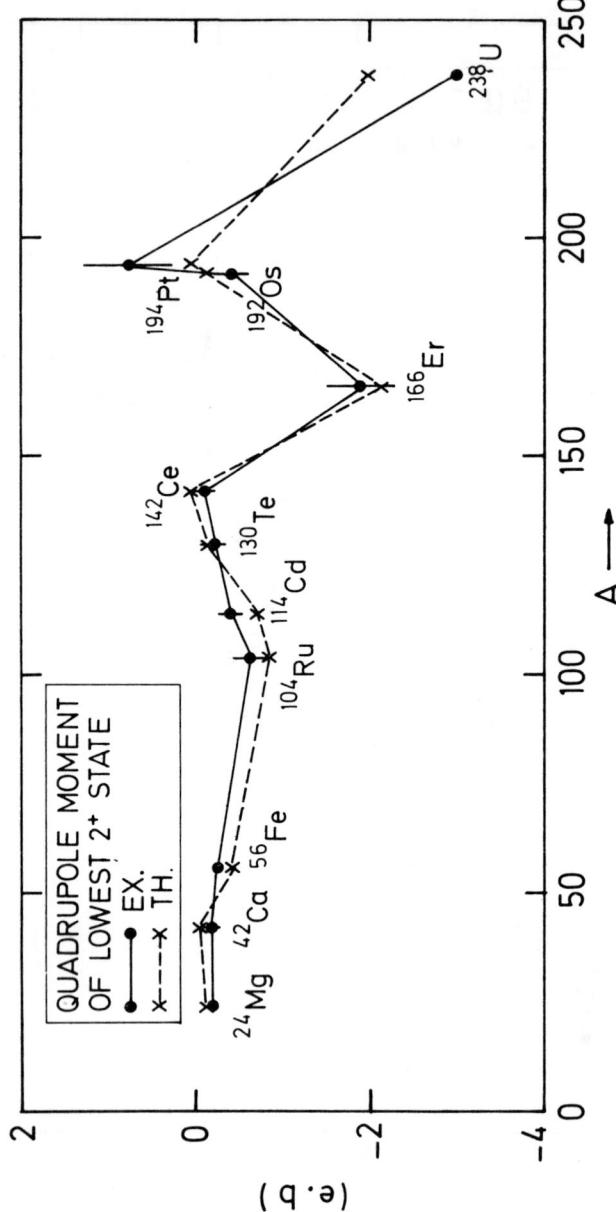

Fig.17. Dynamic Deformation Model (Kumar 1979a).

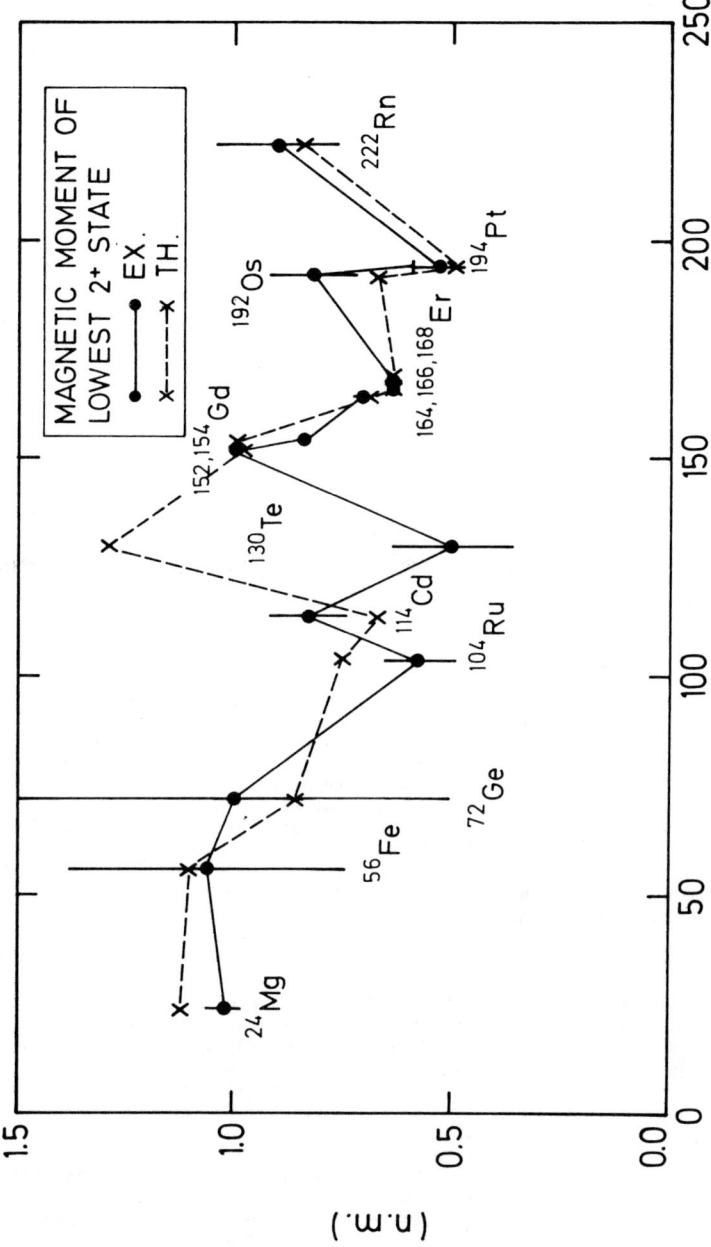

Fig.18. Dynamic Deformation Model (Kumar 1979a).

1. Shape Co-existence in ^{16}O: This example has been discussed extensively by Brown et al. (for example, Brown 1964; Brown and Green 1966). The nucleus ^{16}O is a doubly magic nucleus. Hence, it is expected to be an ideal, spherical nucleus. The ground state ($I^{\pi}=0^+$) is expected to be a 0P-0H state. The lowest even parity states are expected to be of type 2P-2H, since the 1P-1H states would require excitations from the filled N=1(p) shell to the empty N=2 (s-d) shell and, hence, would have negative parity. The lowest energy of 2P-2H excitations (to leading order in an expansion in the interaction strength) is $2\hbar\omega$, where $\hbar\omega$ is the single-particle energy. Using the $\hbar\omega$ value from Eq. (4.8) one finds that the expected excitation energy of the first excited $0'^+$ state is 32.5 MeV. However, the experimental energy is only 6 MeV! Such a tremendous lowering of the $0'^+$ state in an ideal, spherical nucleus caused considerable surprise in the early sixties.

Brown et al. showed that the tremendous lowering of the $0'^+$ state in ^{16}O can be explained by considering this state to be a deformed 2P-2H state rather than a spherical 2P-2H state. They also showed that this concept of shape-coexistence (since the ground state remains spherical) is confirmed by the large B(E2) value connecting the $0'^+$ state to the 2^+ state at 11.08 MeV, the experimental B(E2; $2^+ \to 0'^+$) being 30±5 W.u. (Weisskopf or single-particle units). In contrast the B(E2; $2^+ \to 0^+$) is only 3.4 (3) W.u.

However, Brown et al. had to invoke a number of parameters in order to explain the shape-coexistence in ^{16}O. In contrast, the Dynamic Deformation Model can explain this phenomenon quite easily without invoking any special effects or parameters. Although the calculated B(E2; $2^+ \to 0'^+$) value is only 17 W.u., the calculated ratio B(E2; $2^+ \to 0'^+$)/B(E2; $2^+ \to 0^+$) is even larger than the experimental one. Also, the calculated $0'^+$ is lower than the observed one (see Fig. 19).

The DDM supports the idea of shape co-existence in ^{16}O, but the mechanism is somewhat different from the lowering of 2P-2H states. The calculated potential energy surface (see Fig. 20) exhibits a spherical minimum at $\beta=0$ and a secondary, deformed minimum at $\beta=0.45$, $\gamma=0°$. The deformed minimum is 8.8 MeV above the spherical one. The two minima are separated by a barrier which is 10.4 MeV above the spherical minimum but only 1.6 MeV above the deformed one. It may be tempting to associate the calculated $0'^+$ state at 4.1 MeV with the deformed minimum at $\beta=0.45$, $\gamma=0°$. But this would be incorrect. The $0'^+$ wave function is not localized around the deformation associated with the minimum. Instead, it is a β-vibrational type of wave function with a node at $\beta \sim 0.4$. However, the basic idea of shape-coexistence is supported by the fact that the calculated rms value of β is twice as large for the $0'^+$ state as for the ground state.

This calculation demonstrates very clearly our point of view that even the most "spherical" nuclei can be described within a model which emphasizes nuclear deformations, provided the complete dynamics is taken into account. Naturally, the effects of dynamics are tremendous in such nuclei. In the case of ^{16}O, the $0'^+$ state is lowered from 73 MeV (for harmonic quadrupole vibrations around the spherical minimum) to 4.1 MeV. This is caused partly by the anharmonicity of the potential, but largely it is due to the anharmonicity of the six inertial functions—the three mass parameters for β-, $\beta\gamma$-, γ- vibrations and the three moments of inertia. For instance, the mass parameter B_{00} increases from 0.1 MeV^{-1} at $\beta \sim 0$ to 1.2×10^5 MeV^{-1} at $\beta \sim 0.7$. Such sharp changes are caused by the fact that the single-particle levels move up and down with deformation quite steeply in this region. They cause sharp changes in the single-particle "gap" near the Fermi surface, as well as in the pairing gap. The pairing gap

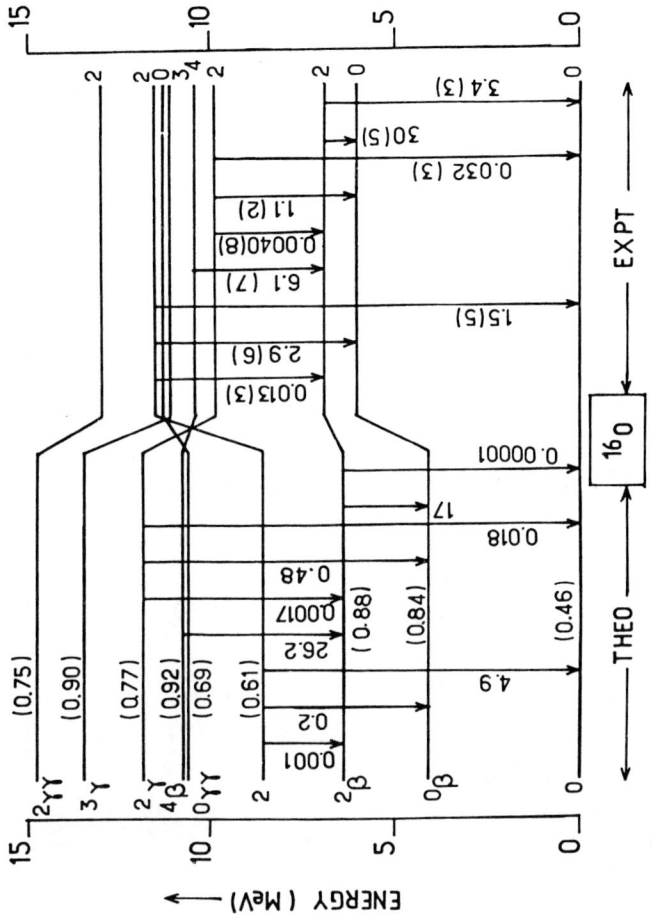

Fig. 19. Low-Energy States of ^{16}O. (Fig. from Kumar 1979b).

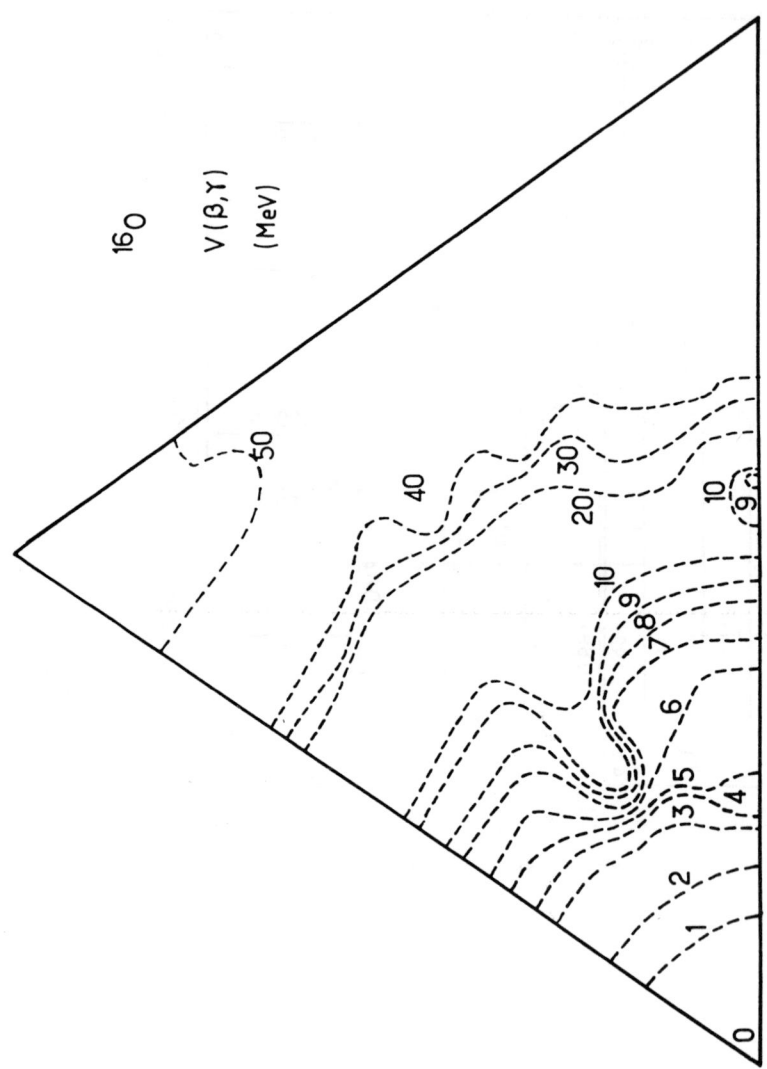

Fig. 20. Contour Plot of the Potential Energy of ^{16}O. (Fig. from Kumar 1979b).

remains zero from $\beta=0$ to $\beta\sim0.5$, but then rises sharply to 1.9 MeV at $\beta\sim0.6$ and then vanishes again at $\beta\sim0.8$. Such sharp variations represent mixings with not only 2P-2H states, but also with 4P-4H, 6P-6H,...

2. Shape-Coexistence in ^{72}Se

The shape coexistence is not quite so dramatic here as compared to ^{16}O. But it is fairly dramatic (also, in Ge isotopes) as compared to most of the collective (non-magic) nuclei where the first excited state is invariably a 2^+ state and the $0'^+$ state usually lies at much higher energy. Instead, the first excited $0'^+$ state of ^{72}Se is only 75 keV above its first 2^+ state. Furthermore, the B(E2) value for the $0'^+ \to 2^+$ transition is 36±7 W.u. (Hamilton et al. 1974).

Hamilton et al. (1974, 1976) interpreted this as evidence for the coexistence of spherical and deformed shapes in ^{72}Se. On the basis of a least-squares fit of a collective Hamiltonian (with 6 fitting parameters for the potential energy and one for the kinetic energy), they deduced that while the ground state is associated with the spherical shape, the $0'^+$ state is associated with a deformed (prolate) potential minimum at $\beta=0.4$.

Some of the results of DDM for ^{72}Se are compared with experiment in Fig. 21 (taken from Kumar 1979b). The figure shows five theoretical bands. Two of these (o, β_o) are associated with the lowest potential minimum (oblate, $\beta=0.41$, $\gamma=48°$, see Fig. 22), two (p, β_p) are associated with the secondary potential minimum (prolate, $\beta=0.43$, $\gamma=6°$, 2 MeV above the lowest minimum), and the fifth one (γ) is common to both minima. The two types of bands are not pure since the barrier between them is only 0.5 MeV above the prolate minimum.

Comparison with experiment shows that the calculated bands are too compressed. This fact, combined with the B(E2) values (see Kumar 1979b), suggests that the calculated minima are too deformed, the oblate one by about 80% and the prolate one by about 50%. However, the general predictions of Hamilton et al. are confirmed in that the $0'^+$ state is deformed (calculated β_{rms} is 0.53) and that it is prolate.

3. Shape Co-existence in ^{240}Pu (Fission Isomers)

Transuranic nuclei exhibit the phenomenon of fission isomers which is related to the existence of a second minimum at about twice the ground state deformation (Strutinsky 1966). Although the term "shape co-existence" is not usually applied to this phenomenon, it falls under the same category in that the same nucleus exhibits two quite distinct shapes under different conditions.

The DDM is not a suitable model for describing the late stages of the fission process where the nucleus starts to break up into two or more fragments. But it appears that it can be employed for describing the potential energy of deformation during the path taken from the ground state deformation, over the inner fission barrier (s) and up to the fission-isomer associated deformations (Kumar 1979b).

While many other models of fission are more advanced from the point of view of the fission degree of freedom (especially the two-centered-shell model developed by Mustafa, Mosel and Schmitt 1973), the DDM is more advanced from the point of view of the dynamics of the early stages of fission.

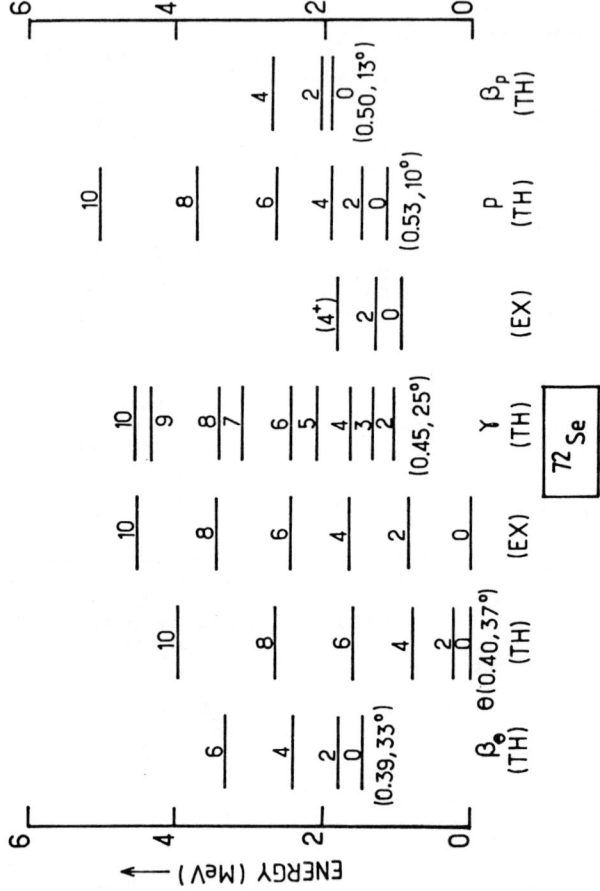

Fig. 21. Low-Energy States of ^{72}Se. (Fig. from Kumar 1979b).

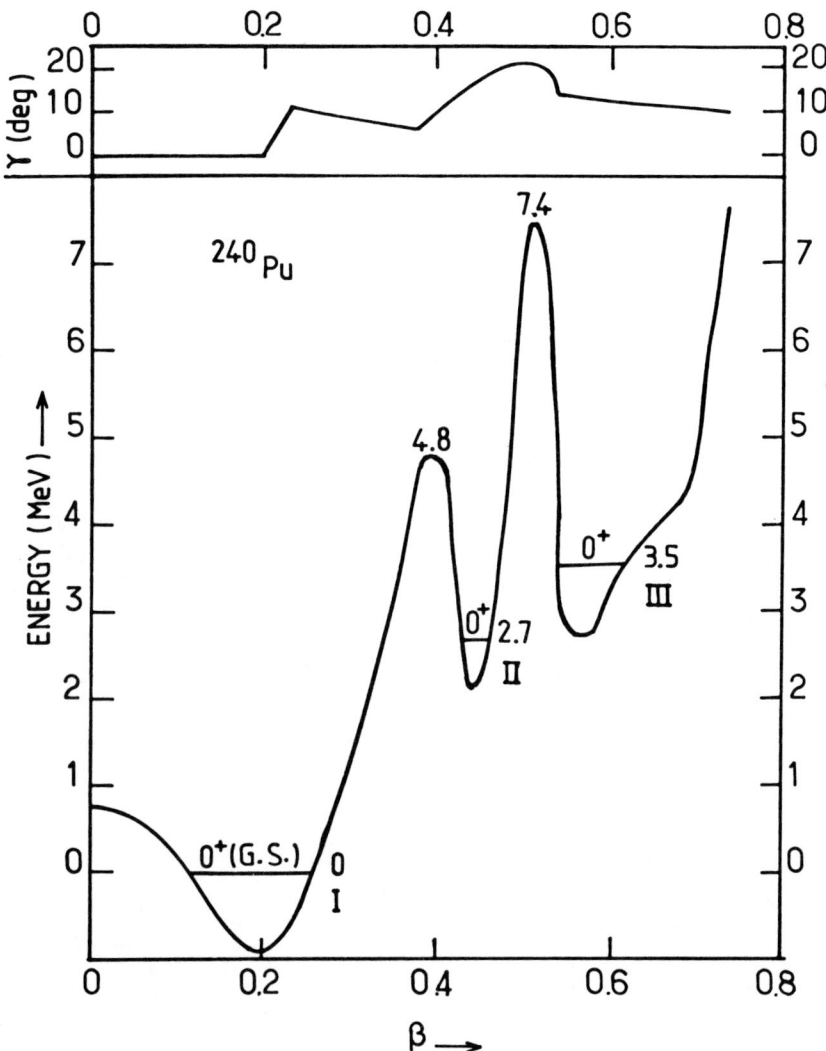

Fig.22. Fission Isomers of ^{240}Pu. The upper part of the figure gives the γ-value along the "fission path" in β. The lower part gives the β-dependence of Veff, which includes the "average" kinetic energy (Kumar 1979b).

The hexadecapole deformation (β_4) included in the DDM is probably too small (see Kumar et al. 1977 for numerical estimates). But the inclusion of dynamics appears to compensate for most of this lack (compared to static models) and the calculated fission barriers of ^{240}Pu are in reasonable agreement with the empirical values (see Fig. 22, taken from Kumar 1979b). What is more remarkable is the agreement with the experimental excitation energies of the fission-isomers: According to the data tables (Ewbank et al. 1979), there are two fission isomers in ^{240}Pu at 2.8 (2) MeV and at 3.0 (2) MeV. The calculated values are 2.7 MeV and 3.5 MeV.

On the other hand, this calculation gave a "moment of inertia" which is too small by a factor of 2.2 for the ground state, and by 1.6 for the fission isomer state. This suggests that the configuration space is not large enough for such heavy nuclei. Calculations with an extended configuration space (N=0-10 major shells instead of N=0-8, about twice as many single-particle levels) are in progress.

D. Rotational - Vibrational Bands in Er nuclei

While ^{16}O represents the spherical limit of collective motion, the Er isotopes represent the deformed limit of collective motion (as far as ground states are concerned). The estimated ground state deformation of the Er nuclei is ~ 0.3, the first 2^+ state occurs at ~ 80 keV, the ratio E_4/E_2 is ~ 3.3. (It reaches a maximum value of 3.31 at ^{168}Er.), the B(E2; 2→0) value is 154 W.u.

Until recently, such nuclei were considered to be among the best understood nuclei (as almost perfect rotors). However, recently Warner et al. (1980) and Davidson et al. (1981) have discovered many new states of ^{168}Er at 1-2 MeV excitation energy which do not fit the ideal rotor picture. They have suggested that the rotor picture should be replaced by the picture of interacting bosons of the model developed by Arima and Iachello (1978), who describe collective motion in terms of interactions among s(J=0) and d(J=2) bosons. However, according to Bohr and Mottelson (1982), the IBM is inconsistent with experiment: A consistent choice of the IBM parameters would give γ-band to g-band transitions which would be several orders of magnitude smaller than those observed for ^{168}Er. On the other hand, Bohr and Mottelson appear to agree with the conclusion that the higher bands (K=0_2, 0_3, 2_2, 2_3, 4_1, 4_2-bands at ≥1.2 MeV) do not fit the ideal rotor picture. In fact, they conclude that the new data imply a static γ-deformation ($\gamma_0 \sim 10°$) for ^{168}Er.

The DDM results presented below (and not published previously) lead to yet another conclusion. Fig. 23 gives a comparison with experiment of the calculated energy levels, B(E2) values, quadrupole moments, and magnetic moments of ^{168}Er. Here we should note once again that no parameters have been varied to fit the data, that no effective charges have been employed,... In view of this, agreement with experiment is quite remarkable (see also Table 8). The major discrepancies concern the ordering of levels in the γγ-band and the γβ-band.

Should one worry that the "normal" ordering of levels is not followed in the calculated γγ- and γβ-bands? Here, we should point out how the theoretical levels are grouped into these bands. These classifications are based on the calculated K-components (see Table 9) and B(E2) values (some of which are given

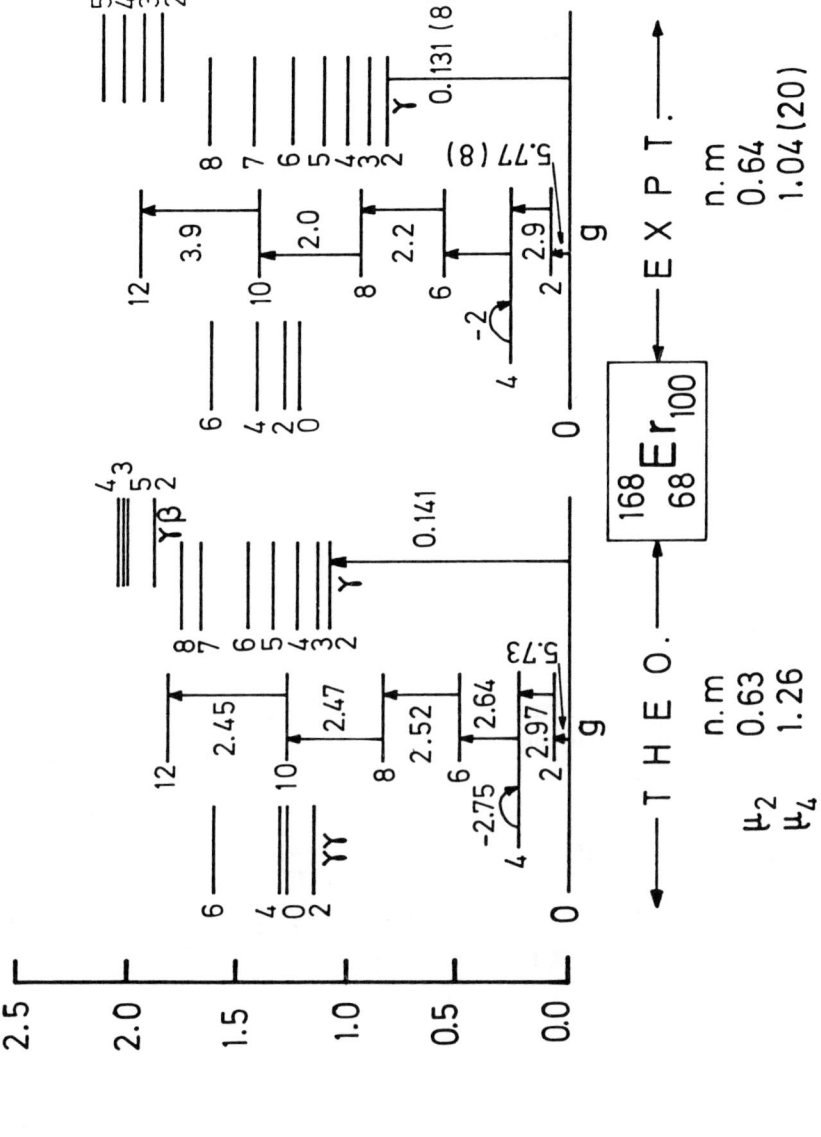

Fig.23. Rotational-Vibrational Bands in ^{168}Er. Experimental data are from Warner et al. (1980) and Greenwood (1974).

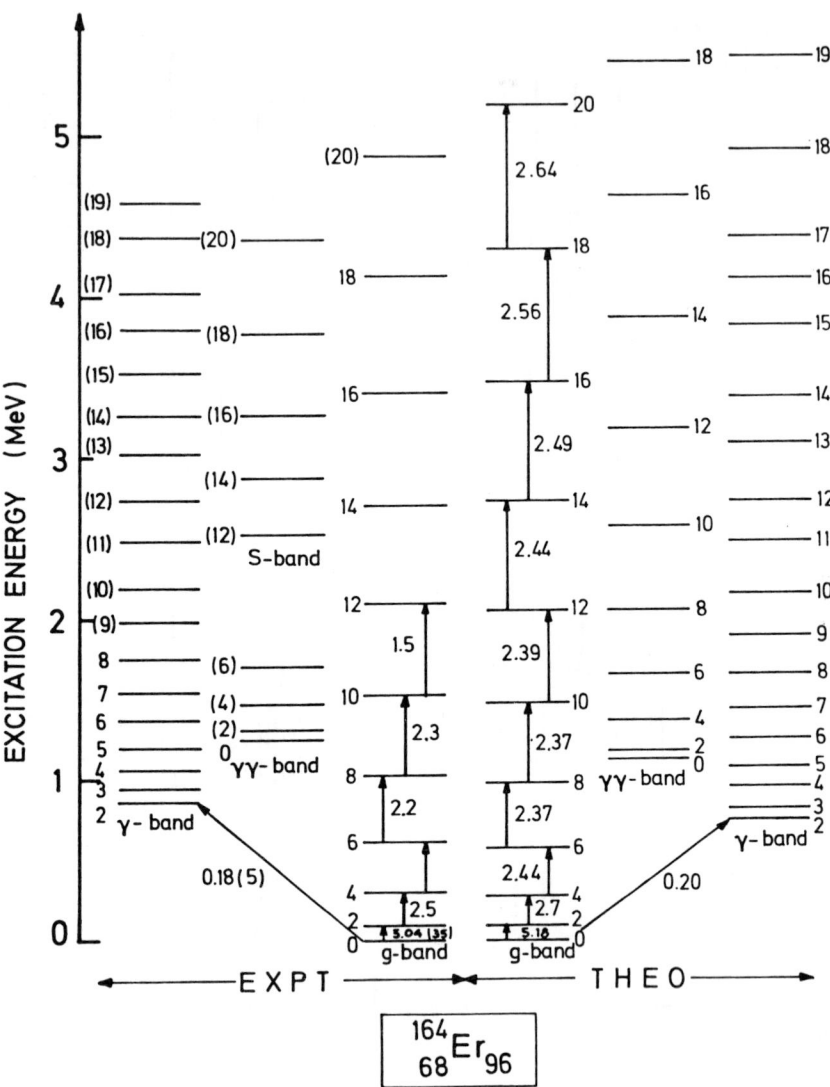

Fig. 24. Rotational-Vibrational Bands in ^{164}Er. Experimental Data are from Johnson et al. (1978) and Buyrn (1974).

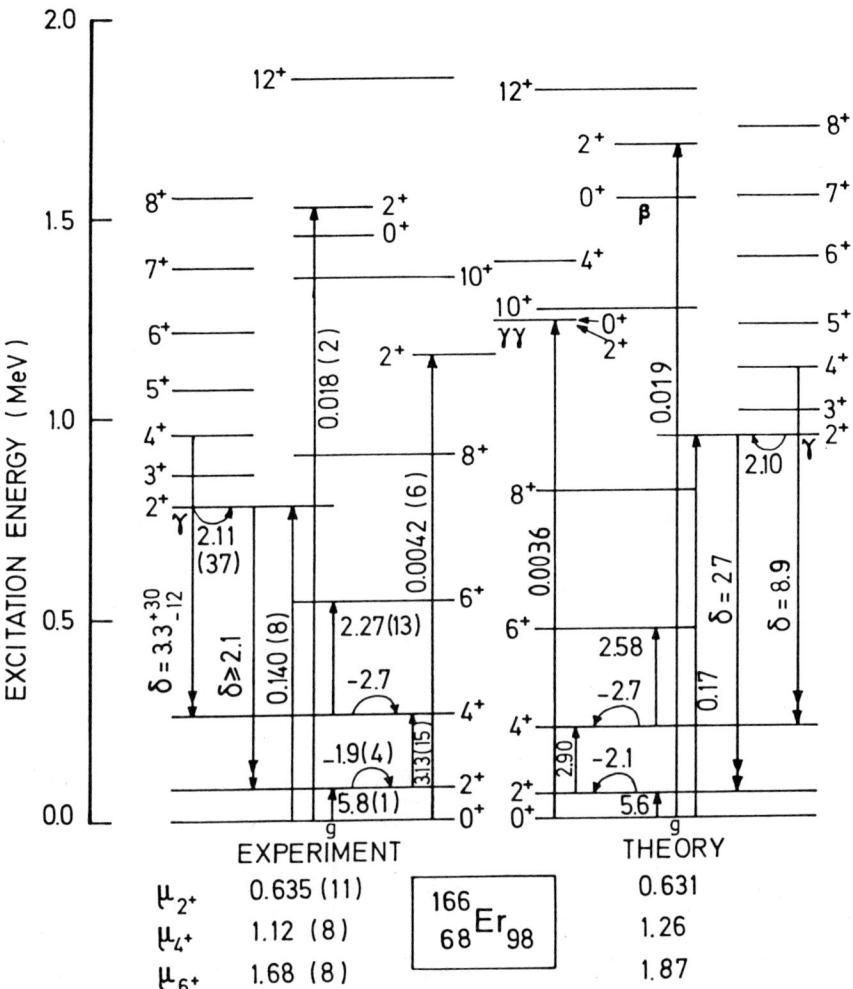

Fig. 25. Rotational-Vibrational Bands in ^{166}Er. The experimental data are from Buyrn (1975) and McGowan et al. (1978).

Table 8. Electromagnetic Moments of ^{164,166,168}Er.

Moment	States	^{164}Er TH	^{164}Er EX	^{166}Er TH	^{166}Er EX	^{168}Er TH	^{168}Er EX
B(E2) (e^2-b^2)	$0_g \to 2_g$	5.18	5.04(35)[a]	5.59	5.8(1)[b]	5.73	5.77(8)[c]
	$0_g \to 2_\gamma$	0.203	0.18(5)[a]	0.171	0.140(8)[d]	0.141	0.131(8)[d]
	$0_g \to 2_\beta$	0.028		0.0036	0.0042(6)[f]	0.032	
	$0_g \to 2_{\gamma\gamma}$	0.001		0.019	0.018(2)[f]	0.00002	
	$2_\gamma \to 0_{\gamma\gamma}$	0.029		0.050		0.058	
	$2_\gamma \to 2_{\gamma\gamma}$	0.027		0.017		0.053	
	$2_g \to 4_g$	2.70	2.5[e]	2.90	3.13(15)[b]	2.97	2.9[e]
	$4_g \to 6_g$	2.44		2.58	2.27(13)[b]	2.64	
	$6_g \to 8_g$	2.37	2.2[e]	2.49	2.4[e]	2.52	2.2[e]
	$8_g \to 10_g$	2.37	2.3[e]	2.46	2.6[e]	2.47	2.0[e]
	$10_g \to 12_g$	2.39	1.5[e]	2.45	2.5[e]	2.45	3.9[e]
Q (e-b)	2_g	-2.04		-2.14	-1.9(4)[b]	-2.17	
	4_g	-2.55		-2.71	-2.7[f]	-2.75	-2[f]
	2_γ	2.04		2.10	2.11(37)[g]	2.12	
μ (n.m.)	2_g	0.69	±0.706(20)[a]	0.63	0.635(11)[b]	0.63	0.64[c]
	4_g	1.38		1.26	1.12(8)[b]	1.26	1.04(20)[c]
	6_g	2.06		1.87	1.68(8)[b]	1.89	
δ (E2/M1)	$2_\gamma \to 2_g$	-2260.		27.	≥21[b]	23.	
	$4_g \to 4_g$	34.		8.9	3.3^{+30}_{-12}[b]	5.7	

[a] Buyrn(1974)
[b] Buyrn(1975)
[c] Greenwood(1974)
[d] McGowan et. al.(1978)
[e] Andrejtscheff et. al.(1975)
[f] Fuller(1976)
[g] McGowan et. al.(1977)

Table 9. Calculated K-Components and other Properties of ^{168}Er.

Band	I	E(MeV) EX[a]	E(MeV) TH	W.F. Component K=0	W.F. Component K=2	W.F. Component K=4	W.F. Component K=6	β rms	γ rms (deg.)	Q (e-b)	μ (n.m.)
0_1	0	0.0	0.0	100				0.315	9	0.0	0.0
(g)	2	0.080	0.055	100	0			0.315	9	-2.17	0.63
	4	0.264	0.224	100	0	0		0.315	8	-2.75	1.26
	6	0.549	0.484	100	0	0	0	0.315	8	-3.01	1.89
	8	0.928	0.829	99	0	0	0	0.316	8	-3.14	2.52
	10	1.396	1.268	99	1	0	0	0.316	8	-3.22	3.14
	12	1.942	1.809	98	2	0	0	0.317	8	-3.26	3.77
2_1	2	0.821	1.075	0	100			0.314	15	2.12	0.68
(γ)	3	0.896	1.127	0	100			0.315	15	0.00	1.00
	4	0.995	1.226	34	65	0		0.318	15	-1.05	1.36
	5	1.118	1.335	0	100	0		0.317	15	-1.70	1.68
	6	1.264	1.451	54	46	0	0	0.321	15	-1.85	2.04
	7	1.433	1.658	0	99	1	0	0.318	15	-2.28	2.36
	8	1.627	1.763	58	41	1	0	0.321	15	-2.08	2.73
	9		2.076	0	96	4	0	0.319	15	-2.50	3.05
0_2	0	1.217	1.266	100				0.318	16	0.00	0.00
(ɱ)	2	1.277	1.150	99	1			0.324	15	-2.21	0.68
	4	1.411	1.301	65	35	0		0.322	14	-2.74	1.34
	6	1.617	1.600	45	54	0	0	0.319	14	-3.17	1.99
	8		1.990	41	58	1	0	0.317	15	-3.43	2.65

Table 9. (continued)

Band	I	E(MeV)		W.F. Component				β	γ rms	Q	μ
		EX[a]	TH	K=0	K=2	K=4	K=6	rms	(deg.)	(e-b)	(n.m.)
2_2	2	1.848	1.874	98	2			0.318	17	-2.05	0.76
(γβ)	3	1.932	2.033	0	100			0.307	20	0.00	1.07
	4	2.022	2.066	89	1	10		0.342	13	-2.30	1.48
	5	2.109	2.021	0	82	18		0.313	17	-0.99	1.73
	6		2.208	97	2	1	0	0.355	9	-3.38	2.19
	7		2.219	0	85	13	2	0.317	16	-1.66	2.40
0_3	0		1.927	100				0.274	24	0.00	0.00
(β)	2		2.062	82	18			0.303	16	-1.01	0.77
	4		2.194	76	22	2		0.295	21	-1.19	1.51
	6		2.384	75	18	6	1	0.303	21	-1.35	2.22
4_1	4	2.030	1.983	10	1	90		0.319	17	3.04	1.33
(γγ)	5		2.099	0	12	88		0.319	17	1.40	1.64
	6		2.302	3	3	94	0	0.315	18	0.36	2.04
	7		2.472	0	4	94	1	0.316	18	-0.41	2.38
0_4	0		2.142	100				0.304	18	0.0	0.0
(ββ)	2		2.251	46	54			0.285	23	0.21	0.77
	4		2.513	64	34	2		0.315	22	-0.96	1.46
	6		2.718	65	28	5	2	0.323	21	-1.38	2.23

[a]Greenwood(1974), Andrejtschaff et al.(1975), Warner et al.(1980)

in Table 8). Note that the K-selection rules of the rotational model apply strictly only to the ground-band here. The g-band members are K=0 (less than 3% admixtures of higher K's for I=2-12), but substantial K-admixtures appear already in the K=2_1 (γ-vibrational) band: The 4+ member is 34% K=0. This is because the 0_2 ($\gamma\gamma$)-band starts at only 0.4 MeV above the 2_1 (γ)-band and both bands have strong admixtures of the pure N_γ=1,2 bands. The 0^+ member of the $\gamma\gamma$-band experiences no mixing since the γ-band has no I=0 member. But the 2^+-member experiences the mixing and is brought down below the 0^+. In fact, the situation is much more complicated than the ideal one of the mixing of two bands only. All the bands are mixed by the Hamiltonian. The Hamiltonian-matrix is diagonalized for each I-value and no separation into bands is assumed at this stage (as in the conventional collective models). Only after the calculation is finished, the calculated states are grouped into bands for the sake of easier identification.

We note here that Fig. 23 does not show some experimental as well as theoretical bands. The experimental bands K=0_3 (1.422 MeV), 2_3 (1.930 MeV), and 4_2 (2.055 MeV) have no counterparts in the calculated spectrum. As shown in Table 9, the next calculated K=0-band occurs at 1.927 MeV and is a β-vibrational band. On the other hand, the observed K=4_1-band (2.030 MeV) is reproduced quite well at 1.983 MeV (see Table 9).

Our conclusion is that although the simple rotational picture breaks down already in the first excited (γ-) band, the Dynamic Deformation Model is valid up to much higher energies. The additional bands must be ascribed to two-quasi-particle bands with higher seniority ($\geqslant$ 2) which are not included in the present version of the model. These conclusions may be tested further by measuring values of many quantities predicted in Tables 8 and 9.

Additional comparisons with the experimental levels and moments of ^{164,166}Er are presented in Fig. 24, 25. In the case of ^{164}Er, we present another test of the DDM. Does this model reproduce the backbending phenomenon observed at high spins (see, for instance, Lieder and Ryde 1978)? The answer is no, since this phenomenon is based on the lowering of the I$\geqslant$ 16 members of the "S-band" below the corresponding members of the g-band. However, if we consider the spectrum as a whole, we have to conclude that the present version of the DDM is reasonable up to quite high spins. Incidentally, the solution of the collective Schrödinger equation up to I=20 required some major improvements in the "Complete Numerical Solution" method (as discussed in ch. III) as compared to the original version of the method (Kumar 1963, Kumar and Baranger 1967) which was limited to I_{max}=4 because of numerical limitations.

On the other hand, the backbending phenomenon is an important phenomenon. Its explanation, in the context of the DDM, would require the inclusion of two quasiparticle states having seniority quantum number equal to two.

X. Conclusions

The main conclusion of the present study of collective motion in atomic nuclei is that nuclei are quite collective and deformable--not only the heavy, non-magic nuclei like the ^{164,166,168}Er nuclei but also the light, doubly-magic nuclei like ^{16}O. As Gerry Brown said already in 1964, "All nuclei are deformed." In the language of the DDM, the rms value of β is non-zero for all nuclei. Also, the rms value of γ is non-zero for all nuclei.

The forgoing statements are based not only on the numerical results of the Dynamic Deformation Model discussed above. They are also based on a general theory of moments which relates the rms values of β and γ directly to the observable B(E2) values in a model-independent way (Kumar 1972, 1975a).

Furthermore, a nucleus does not have a fixed shape. The values of β_{rms}, γ_{rms} are different for different states of the same nucleus. In some cases, they do not change much--as in the ground state rotational bands of well-deformed nuclei like ^{168}Er. In other cases (e.g. ^{16}O, ^{72}Se, ^{240}Pu), they change so dramatically that the phenomenon is called "shape co-existence".

If all nuclei are deformed and if the shape is different for different nuclear states, why is the concept of "nuclear shape" interesting? We offer the following arguments.

1. It provides a simple, picturesque way of understanding and unifying many aspects of nuclear phsyics. Instead of only comparing a table of calculated numbers with another table of experimental numbers, the "deformed" models provide a potential energy of deformation. The lowest minimum of this potential can be employed to make some general statements about the low-energy spectra. Extension of this potential to large enough deformations allows us to unify the theories of nuclear structure and nuclear fission. This unification is not only pleasing from an aesthetic point of view. It also has a practical aspect: The correct behaviour of the potential at large deformations is essential for the correct behaviour of nuclear wave functions.

2. The extension of the concept of the potential energy of deformations to the solution of the full Schrödinger equation of deformations, as implemented in the Dynamic Deformation Model, allows us to unify (a) spherical-transitional-deformed even-even nuclei, and (b) light-medium-heavy even-even nuclei.

3. The concept of deformation has been found to be quite useful also in the theory of odd-A nuclei. However, quite different models have been employed in the past for "spherical" and for "deformed" odd-A nuclei. Their unification via DDM is in progress as follows. Very briefly, the DDM wave functions for even-even cores (A-1 and A+1) are coupled to a particle or a hole to describe an odd-A nucleus. The two sets of wave functions are further coupled to each other via a pairing-type interaction in a Kerman-Klein (Dreiss et al. 1971) type of approach. The "physical" states out of the doubled set of states are chosen according to the criterion developed by Dönau and Frauendorf (1977). A preliminary calculation for ^{191}Pt (Vaagnes,

Kumar, and Vaagen 1982) has been quite successful, but much more work needs to be performed in order to achieve a satisfactory unification of spherical and deformed odd-A nuclei.

4. The concept of deformation has been found to be quite useful also in the theory of nuclear reactions. In the coupled channel approach, pioneered by Tamura (1963), the coupling between different reaction channels is determined by the deformation of the target nucleus. The DDM wave functions have recently been employed to unify the "spherical" (vibrational) and the "deformed" (rotational) versions of the coupled channel approach. The coupling potentials are calculated for different points of the β-γ mesh and then weighted according to the DDM wave functions. Preliminary calculations for (n, Sm) cross-sections (Lagrange, Girod, Grammaticos, and Kumar 1980), and for (p, Ge) cross-sections have recently been reported. Thus, another unification (this one of nuclear structure and nuclear reaction theories) is in progress.

This has been a preliminary study whose main objective was to determine the feasibility of attempting to unify the various aspects of nuclear theory. Our main conclusion is that it is quite feasible to do so. But from the point of view of making the theory more accurate (and more complete in some instances, for example odd-parity states in e-e nuclei), it is clearly necessary to include two quasi-particle states more explicitly in the Dynamic Deformation Model.

BIBLIOGRAPHY

Aase, A., K. Kumar, and J. S. Vaagen, 1980, in Proceedings International Conference on Nuclear Physics, Berkeley, August 1980 (University of California, Berkeley) Vol. 1, p. 305.

Ajzenberg-Selove, F., 1971, Nucl. Phys. A281, 1.

Alaga, G., K. Alder, A. Bohr, and B. R. Mottelson, 1955, Mat. Fys. Medd. Dan. Vid. Selsk. 29, No. 9.

Alder, K., and Aa. Winther, 1966, Coulomb Excitation (Academic Press, New York).

Andrejtscheff, W., K. D. Schilling, and P. Manfrass, 1975, At. Data Nucl. Data Tables 16, 515.

Ardouin, D., B. Remaud, K. Kumar, F. Guilbault, P. Avignon, R. Selz, M. Vergnes, and G. Rotbard, 1978, Phys. Rev. C18, 2739.

Baranger, M., 1963, in Cargese Lectures in Theoretical Physics (Benjamin, New York).

Bardeen, J., L. N. Cooper, and J. R. Schrieffer, 1957, Phys. Rev. 106, 162.

Belyaev, S. T., 1959, Mat. Fys. Medd. Dan. Vid. Selsk. 31, No. 11.

Bes, D. R., R. A. Broglia, R. P. J. Perazzo, and K. Kumar, 1970, Nucl. Phys. A143, 1.

Bogolyubov, N. N., 1958, Nuovo Cimento, 7, 794.

Bohr, A., 1952, Mat. Fys. Medd. Dan. Vid. Selsk., 26, No. 14.

Bohr, A., and B. R. Mottelson, 1953, Mat. Fys. Medd. Dan. Vid. Selsk., 27, No. 16.

Bohr, A., and B. R. Mottelson, 1969, Nuclear Structure (Benjamin, New York).

Bohr, A., and B. R. Mottelson, 1975, Nuclear Structure (Benjamin, New York).

Bohr, A., B. R. Mottelson, and D. Pines, 1958, Phys. Rev. 110, 936.

Bohr, A., and B. R. Mottelson, 1982, Phys. Scr. 25, 28.

Brack, M., J. Damgaard, H. C. Pauli, A. S. Jensen, V. M. Strutinsky, and C. Y. Wong, 1972, Rev. Mod. Phys. 44, 42.

Brown, G. E., 1964, Proc. Int. Conf. Nucl. Phys. (CNRS, Paris, 1964), p. 129.

Brown, G. E., 1967, Unified Theory of Nuclear Models (North-Holland, Amsterdam).

Brown, G. E., and A. M. Green, 1966, Nucl. Phys. 85, 87.

Buyrn, A., 1974, Nucl. Data Sheets 11, 327.

Buyrn, A., 1975, Nucl. Data Sheets 14, 471.

Casimir, H. B. G., 1936, On the Interaction Between Atomic Nuclei and Electrons
(Prize Essay, Teyler's, Tweede, Haarlem).

Davydov, A. S., and A. A. Chaban, 1960, Nucl. Phys. 20, 499.

de-Shalit, A., and H. Feshbach, 1974, Theoretical Nuclear Physics (Wiley, New York).

Dobaczewski, J., S. G. Rohozinsky, and J. Srebrny, 1975, Nukleonika 20, 981.

Dreiss, G. J., R. M. Dreizler, A. Klein, and G. Do Dang, 1971, Phys. Rev. C3, 2412 and references quoted there.

Dönau, F., and S. Frauendorf, 1977, Phys. Lett. 71B, 263.

Elliott, J. P., 1958, Proc. Roy. Soc. (London) A245, 128; 562.

Ewbank, W. B., Y. A. Ellis, and M. R. Schmorak, 1979, Nucl. Data Sheets 26, 1.

Fuller, G. H., 1976, J. Phys. Chem. Ref. Data 5, 835.

Giraud, B., and B. Grammaticos, 1974, Nucl. Phys. A233, 373.

Glenn, J. E., and J. X. Saladin, 1968, Phys. Rev. Lett. 20, 1298.

Goeke, K., and P. G. Reinhard, Ann. Phys. (N.Y.) 112, 328.

Greenwood, L. R., 1974, Nucl. Data Sheets 11, 385.

Greiner, W., 1966, Nucl. Phys. 80, 417.

Gustafson, C., I. L. Lamm, B. Nilsson, and S. G. Nilsson, 1967, Arkiv Fysik 36, 613.

Hamilton, W. D., 1969, Nucl. Phys. A136, 257.

Hamilton, J. H., 1972, Izv. Akad. Nauk, SSSR Ser. Fiz. 36, 17.

Hamilton, J. H., A. V. Ramayya, W. T. Pinkston, R. M. Ronningen, G. Garcia-Bermudez, H. K. Carter, R. L. Robinson, H. J. Kim, and R. O. Sayer, 1974, Phys. Rev. Lett. 32, 239.

Hamilton, J. H., H. L. Crowell, R. L. Robinson, A. V. Ramayya, W. E. Collins, R. M. Ronningen, V. Maruhn-Rezwani, J. A. Maruhn, N. C. Singhal, H. J. Kim, R. O. Sayer, T. Magee, and L. C. Whitlock, 1976, Phys. Rev. Lett. 36, 340.

Hamilton, W. D., and K. E. Davies, 1968, Nucl. Phys. A122, 165.

Hasse, R. W., 1971, Ann. Phys. (N.Y.) 68, 377.

Haxel, O., J. H. D. Jensen, and H. E. Suess, 1949, Phys. Rev. 75, 1766.

Hill, D. L., and J. A. Wheeler, 1953, Phys. Rev. 89, 1102.

Inglis, D. R., 1955, Phys. Rev. 97, 701.

Johnson, N. R., D. Cline, S. W. Yates, F. S. Stephens, L. L. Riedinger, and R. M. Ronningen, 1978, Phys. Rev. Lett. 40, 151.

Kishimoto, T., and T. Tamura, 1976, Nucl. Phys. A270, 317.

Kisslinger, L. S., and R. A. Sorensen, 1960, Mat. Fys. Medd. Dan. Vid. Selsk., 32, No. 9.

Krane, K. S., and R. M. Steffen, 1971, Phys. Rev. C3, 240.

Kumar, K., 1963, A Study of Nuclear Deformations with Pairing Plus Quadrupole Forces (Carnegie-Hunt Library, Pittsburgh). Ph. D. Thesis.

Kumar, K., 1972, Phys. Rev. Lett. 28, 249.

Kumar, K., 1974, Nucl. Phys. A231, 189.

Kumar, K., 1975a, in The Electromagnetic Interaction In Nuclear Spectroscopy, ed. W. D. Hamilton (North-Holland, Amsterdam) p. 55; p. 119.

Kumar, K., 1975b, Nukleonika, 20, 133.

Kumar, K., 1975c, Physica Scripta 11, 179.

Kumar, K., 1978, J. Phys. G (London) 4, 849.

Kumar, K., 1979a, in Structure of Medium Heavy Nuclei 1979 (Institute of Physics, London, Conference Series No. 49) p. 169.

Kumar, K., 1979b, in Proceedings International Symposium on Future Directions in Studies of Nuclei Far from Stability, Nashville, September 1979, ed. J. H. Hamilton et al. (North-Holland, Amsterdam) p. 265.

Kumar, K., and Baranger, M., 1966, Phys. Rev. Lett. 17, 1146.

Kumar, K., and M. Baranger, 1967, Nucl. Phys. A92, 608.

Kumar, K., and M. Baranger, 1968, Nucl. Phys. A122, 273.

Kumar, K., B. Remaud, P. Aguer, J. S. Vaagen, A. C. Rester, R. Foucher, and J. H. Hamilton, 1977, Phys. Rev. C16, 1235.

Lagrange, Ch., M. Girod, B. Grammaticos, and K. Kumar, 1980, in Proceedings International Conference on Nuclear Physics, Berkeley, August 1980 (University of California, Berkeley) Vol. 1, p. 352.

Lieb, K. P., and J. J. Kolata, 1977, Phys. Rev. C15, 939.

Lieder, R. M., and H. Ryde, 1978, in Advances in Nuclear Physics, ed. M. Baranger and E. Vogt (Benjamin, New York) Vol. 10, p. 1.

Lien, J. R., J. S. Vaagen, and A. Graue, 1975, Nucl. Phys. A253, 165.

Mayer, M. G., 1949, Phys. Rev. 75, 1969.

Mayer, M. G., 1950, Phys. Rev. 78, 16.

McGowan, F. K., W. T. Milner, R. O. Sayer, R. L. Robinson, and P. H. Stelson, 1977, Nucl. Phys. A289, 253.

McGowan, F. K., W. T. Milner, R. L. Robinson, and P. H. Stelson, 1978, Nucl. Phys. A297, 51.

McGrory, J. B., and B. Wildenthal, 1980, Ann. Rev. Nucl. Part. Sci. 30, 383.

Messiah, A., 1962, Quantum Mechanics (North-Holland, Amsterdam).

Milner, W. T., F. K. McGowan, R. L. Robinson, and P. H. Stelson, 1971, Nucl. Phys. A177, 1.

Mustafa, M. G., U. Mosel, and H. W. Schmitt, 1973, Phys. Rev. C7, 1519.

Myers, W. D., 1976, At. Data Nucl. Data Tables 17, 411.

Myers, W. D., and W. J. Swiatecki, 1974, Ann. Phys. (N.Y.) 84, 186.

Newton, T. D., 1960, Can. J. Phys. 38, 700; Atomic Energy of Canada Limited Report No. CRT-886, 1960.

Nilsson, S. G., 1955, Mat. Fys. Medd. Dan. Vid. Selsk. 29, No. 16.

Nix, J. R., 1972, Ann. Rev. Nucl. Sc. 22, 65.

Piercey, J. B., 1979, pr. comm.

Preston, M. A., and R. K. Bhaduri, 1975, Structure of the Nucleus (Addison-Wesley, Reading).

Pryor, R. J., and J. X. Saladin, 1970, Phys. Rev. C1, 1573.

Racah, G., 1950, Phys. Rev. 78, 622.

Rainwater, J., 1950, Phys. Rev. 79, 432.

Remaud, B., 1978, pr. comm.; Remaud, B., and G. Royer, 1981, J. Phys. A (London) 14, 2897.

Stelson, P. H., and L. Grodzins, 1965, Nucl. Data A1, 21.

Strutinsky, V. M., 1966, Yad. Fiz. 3, 614 [Sov. J. Nucl. Phys. 3, 449].

Sørensen, B., 1970, Nucl. Phys. A142, 411.

Tamura, T., 1963, Rev. Mod. Phys. 37, 679.

Tanabe, K., and K. Sugawara-Tanabe, 1976, Phys. Rev. C14, 1963.

Turner, R. J., and T. Kishimoto, 1973, Nucl. Phys. A217, 317.

von Bernus, L., W. Greiner, V. Rezwani, W. Scheid, U. Schneider, M. Sedlmayr, and R. Sedlmayr, 1974, Proc. Conf. on Problems of Vibrational Nuclei (North-Holland, Amsterdam) p. 230.

Vaagnes, A., K. Kumar, and J. S. Vaagen, 1982, Physica Scripta 25, 443.

Warner, D. D., R. F. Casten, and W. F. Davidson, 1980, Phys. Rev. Lett. 45, 1761; and pr. comm.

Wong, C. Y., 1968, Nucl. Data Tables A4, 271.

Yariv, Y., T. Ledergerber, and H. C. Pauli, 1976, Z. Phys. A278, 225.